The Functional Foods Revolution

The Functional Foods Revolution

Healthy people, healthy profits?

Michael Heasman and Julian Mellentin

Earthscan Publications Ltd, London and Sterling, VA

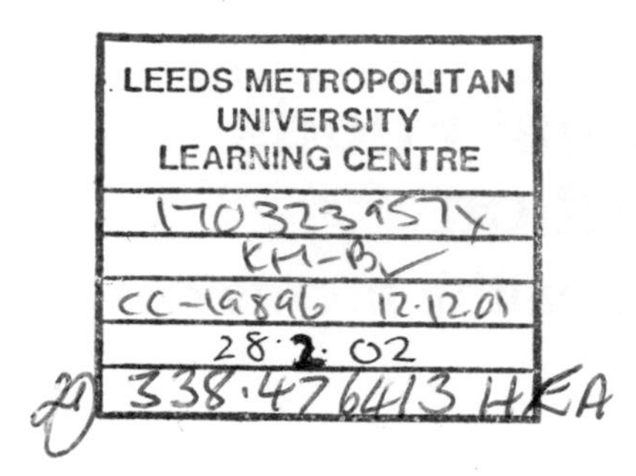

First published in the UK and USA in 2001
by Earthscan Publications Ltd

A catalogue record for this book is available from the British Library

ISBN: 1 85383 687 7 paperback
1 85383 688 5 hardback

Typesetting by PCS Mapping & DTP, Newcastle upon Tyne
Printed and bound by Creative Print and Design Wales, Ebbw Vale
Cover design by Susanne Harris

For a full list of publications please contact:
Earthscan Publications Ltd
120 Pentonville Road, London, N1 9JN, UK
Tel: +44 (0)20 7278 0433
Fax: +44 (0)20 7278 1142
Email: earthinfo@earthscan.co.uk
http://www.earthscan.co.uk

22883 Quicksilver Drive, Sterling, VA 20166–2012, USA

Earthscan is an editorially independent subsidiary of Kogan Page Ltd and publishes in association with WWF-UK and the International Institute for Environment and Development

This book is printed on elemental chlorine-free paper

Contents

Part 5 The Healthful Company™: A Leap Forward

Tables, Figures, Boxes and Photos

Tables

Figures

Boxes

Photos

Acronyms and Abbreviations

ACNFP	Advisory Committee on Novel Foods and Processes
ADA	American Dietetics Association
ANZFA	Australian and New Zealand Food Authority
ARR	absolute risk reduction
ASA	Advertising Standards Authority
BACC	British Advertising Clearance Centre
BEUC	Bureau European des Union de Consommateurs
BMJ	British Medical Journal
BSE	bovine spongiform encephalopathy
CA	Consumers' Association
CAB	Code Administration Body
CAP	Common Agricultural Policy
CFU	colony forming units
CIAA	European Confederation of Food and Drink Industries
COMA	Committee on Medical Aspects of Food Policy
CPG	consumer packaged goods
DHSS	Department of Health and Social Security
DRT	Les Dérivés Résiniques et Terpéniques
DSHEA	Dietary Supplement Health and Education Act
EC	European Commission
EFLA	European Food Law Association
EHPM	European Health Product Manufacturers' Association
EU	European Union
FAC	Food Advisory Committee
FAO	The Food and Agriculture Organization
FBI	Finnish Bread Information
FDA	Food and Drug Administration
FDAMA	Food and Drug Administration Modernization Act
FHP	Functional Foods for Health Program
FIAA	Federation of Agriculture and Food Industries (Belgium)
FMI	Food Marketing Institute
FOSHU	foods for specified health uses
GM	genetically modified
GRAS	generally recognized as safe
HDL	High Density Lipoprotein
IACFO	International Association of Consumer Food Organizations

IFIC	International Food Information Council
IFT	Institute of Food Technology
ILSI	International Life Sciences Institute
IQ	Intelligent Quisine
IRI	Information Resources Inc
IT	information technology
JAMA	*Journal of the American Medical Association*
JHCI	Joint Health Claims Initiative
JHFNFA	Japan Health Food & Nutrition Food Association
LAB	lactic acid bacteria
LACOTS	Local Authority Coordinating Body on Trading Standards
LDL	Low Density Lipoprotein
LGG	Lactobacillus Goldin and Gorbach
LOHAS	lifestyles which combine health and sustainability
MAFF	Ministry of Agriculture Fisheries and Food
NACNE	National Advisory Committee on Nutrition Education
NBJ	*Nutrition Business Journal*
NCC	National Consumer Council
NCI	National Cancer Institute
NFA	National Food Alliance
NGO	non-governmental organization
NLEA	Nutrition Labeling and Education Act
NMHA	National Mental Health Association
NOP	National Opinion Poll
OECD	Organization for Economic Cooperation and Development
RDA	recommended daily allowance
RISC	International Research Institute on Social Change
RRR	relative risk reduction
TEKES	Finland's government-funded Technology Development Centre
TNO	Dutch research institute
TRC	Tomato Research Council
UNDP	United Nations Development Programme
UNICEF	United Nations Children's Fund
USDA	United States Department of Agriculture
VTT	Technical Research Centre of Finland
WHO	World Health Organization

Acknowledgements

This book draws on more than five years of research into food and health by the authors. The manuscript was drafted in our busy office at the Centre for Food & Health Studies, London, fitting around our normal workload.

Our network of researchers around the world have been invaluable in helping put this book together and we acknowledge, too, the efforts of our administrative staff, Miranda Mills, Satwant Sandhu and Julie Eagleton, in helping to manage and edit the mass of material we had to deal with and in helping keep everything under control. Thanks, too, to Kevin Bender of Information Resources Inc., in Chicago, Tom Vierhile of Marketing Intelligence Service, New York, and David Jago at Mintel, London, for their very efficient and speedy help in gathering product information.

Crucially, the foundations of this book lie in the dozens of interviews that we and our team have carried out over the years for *New Nutrition Business*, our monthly journal and we would therefore like to thank the following people for giving us some time from their busy schedules:

Tage Affertsholt, 3A Consulting, Denmark

Tor Bergman, CEO, Raisio Group, Finland

Dr Patrizio Cagnasso, Head of Research, Parmalat, Italy

Bruce Campbell, Marketing Manager, Mainland Products, New Zealand

Dr Iain Cloughley, Managing Director, Clover Healthcare, UK

Professor Kevin Collins, Department of Medicine & Microbiology, National Food Biotechnology Centre, Ireland

Paul Coussement, Marketing Director, Orafti Active Food Ingredients, Belgium

Larry H Cunningham, Vice President, Archer Daniels Midland, USA

Tony De Lio, Vice President, Nutraceuticals Division, Archer Daniels Midland, USA

Dr Nigel Dickie, Nutrition Consultant, Heinz, UK

Anne Duffy, Public Relations Manager, Yakult, Australia

Dr Anne Franck, Application and Nutrition Director, Orafti Active Food Ingredients, Belgium

Craig Fuller, Director, Chilled Dairy, Nestlé, Australia

Ann-Laure Gassin, Nutrition Affairs Manager Europe, Kellogg, France

Jo Goossens, Director, Health & Nutrition Group, Eridania Beghin-Say, Belgium

Professor Dennis Gordon, University of North Dakota, USA

Dr Clare Hasler, Executive Director, Functional Foods for Health Program, University of Illinois, USA

Sten von Hellens, Head of Corporate Communications, Raisio Group, Finland

Hirokatsu Hirano, Managing Director, Yakult Honsha, Japan

Dr Lars Hoie, R&D Director, NutriPharma, Norway

Kalus Holmsberg, Head of Functional Foods, Aarhus Oliefabrik, Denmark

Marika Ingman, Marketing Director, Ingman Foods, Finland

Melvyn Jay, Head of Functional Foods Europe, Novartis Consumer Health, Switzerland

Kalle Laporanta, LGG Project Manager, Valio, Finland

Yvonne Lo, President, Vitasoy, USA

Ron Martell, Vice President of Marketing, Pacific Foods, USA

Brian Maxwell, CEO, PowerBar, USA

Stefano Meloni, President, Eridania Beghin-Say, France

Jim Munday, Public Relations Manager, Yakult, UK

Rickard Oste, CEO, CEBA, Sweden

Al Piergallini, CEO, Novartis Consumer Health Worldwide, Switzerland

Lars Poulson, Manager Functional Foods, MD Foods, Denmark

Professor Kaisa Pountanen, VTT Biotechnology, Finland

Mathew Richardson, Managing Director, Pro-Fibe Nutrition, UK

Dr David Richardson, Head of Science & Comunications, Nestlé, UK

Lars Sjostrand, CEO, BioDoc, Sweden

Mike Smolyansky, CEO, Lifeway Foods, USA

Michael Sneed, Vice President, McNeil Consumer Nutritionals, USA

Steve Sterling, Vice President of Sales & Marketing, Functional Foods, ConAgra, USA

James Taylor, Vice President of Marketing, Dannon, USA

Dr Jon Vanderhoof, Vice President of Nutrition & Health Sciences, ConAgra, USA

Kaj Vareman, CEO, Probi AB, Sweden

Dionne Vernon, Brand Manager, Campbell Soup Co., USA

Carol Wham, Nutrition Marketing Manager, New Zealand Dairy Foods, New Zealand

Rhonda Whitwer, Director of Business Development, GalaGen, USA

Dorothy Wright, Project Manager – Nutrition, Nestlé, UK

Dr Paul Yeung, Professor of Nutrition, Department of Nutrition Sciences, Faculty of Medicine, University of Toronto & Heinz Director of Corporate Nutrition, Canada

About the Authors

Heasman and Mellentin are two of the world's leading commentators on and analysts of 'functional foods'. They are Directors of The Centre for Food & Health Studies, a London-based think-tank that researches, analyses and forecasts developments in food and health around the world, as well as providing policy and strategic advice for companies and consumer organizations. The Centre publishes *New Nutrition Business*, the long-established international journal on food and health and the website www.new-nutrition.com

Dr Michael Heasman has a 13-year record of academic research in food policy and has held research posts at Reading, London and Bradford universities. He is Affiliate Faculty Member at the Functional Foods for Health Program, University of Illinois, Chicago, USA and Visiting Research Fellow at the Centre for Food Policy, Wolfson Institute of Health Sciences, London. Prior to setting up The Centre for Food & Health Studies he was Senior Research Fellow at the Centre for Food Policy, Wolfson Institute of Health Sciences, the leading food policy research unit in the UK, for four years. His advice on food and health is sought by many household name companies, as well as consumer organizations and policy makers. He obtained his PhD at Bradford University in the Food Policy Research Unit.

Julian Mellentin has a 12-year track record in sales and marketing of branded products. For three years he was sales and marketing director of the niche branded products subsidiary of a multinational company, based in Amsterdam and operating across Europe and the Middle East, setting up pan-European distribution and sales and taking the company's market share to 38 per cent within three years. He has lived and worked in The Netherlands, France, Germany and Hong Kong. He obtained his first degree at Oxford University and his MBA at Manchester Business School.

Introduction

A New Era in Food and Health Marketing

The health-promoting effects of foods is the food industry's 'big idea' at the start of the 21st century. Most of the world's major food companies are looking to boost their nutritional and scientific expertise in pursuit of the health benefits of food and introduce products and marketing campaigns that tell people how to beat disease and illness through food consumption. The future of food will increasingly be about how it affects our health and well-being and the sorts of products and ingredients that will deliver such health benefits. This big idea has been termed 'functional food', and this book is the story of how the functional foods 'revolution' is being unleashed across the globe.

One consequence of the functional foods revolution is that the marketing of food, nutrition and health is taking place on an unprecedented scale. But food is more than this: it is essential to our biological existence and, as living material cuts to the beating heart of our culture, personal interactions, and our sense of well-being – after all, 'we are what we eat'. It is little wonder that any real or potential changes to our food supply quickly become embroiled in, for the want of a better term, the politics of food. Functional food is proving to be no exception to this rule.

Developments in functional food have quickly moved to centre stage in public discussions concerning food and health policy. Over the past 20 years there have been major fights over food and public health on numerous battlegrounds – from chemicals and additives used in food production to major food safety crises such as bovine spongiform encephalopathy (BSE) or 'mad cow disease' in Europe. Continuing debate in the developed world focuses on diet-related disease and illnesses, in particular heart disease, certain cancers and obesity. The amounts of total fat (particularly saturated fats in the form of dairy and meat products), sugars, and other nutrients in the diet and their negative or positive roles in the cause or prevention of the diseases of affluence have been the subjects of bitter public controversies.

Food industries have responded in a variety of ways: those negatively affected by public health advice have exhibited denial, arguing that there are no good or bad foods, but only good or bad diets, and placing the responsibility for a good diet on the individual. Another food industry tactic has been to attack the scientific validity of research and even the credibility of individuals who have been vocal critics of what some have called the 'toxic food environment' (Colles, 1998). A further response has been to adapt, by creating a massive new market for healthy eating products, notably sugar-free, fat-reduced and low-fat products.

Now the market for food in relation to health is once again being turned on its head. During the late 1990s the food industry has enthusiastically, almost evangelically, come to embrace the whole new concept of functional foods – foods and beverages that may provide health benefits beyond basic nutrition, and which have been termed 'nutraceuticals' in the US. Many scientists describe the developments in functional food science as standing 'at the threshold of a new frontier in nutritional sciences' (Bellisle et al, 1998). Functional food science is already being actively commercialized and promoted to health-conscious consumers around the globe. The world's top food companies have embraced the functional food concept and see it as a trend which is here to stay.

Fuelling Corporate and Scientific Ambition

After years of negative talk about food in relation to diet and health, functional ingredients are now being used as positive attributes to create new markets. Cutting edge technologies and scientific findings about dietary components promise high value-added products, which can be targeted at niche markets. Corporate investment and market ambition in this area is matched only by the ambition of the functional food science community in its race to unravel the health-enhancing components of foods.

A seemingly endless number of business analysts estimate the potential vastness of the market and value of functional foods. They are less keen, however, to document the number of product failures, the multitude of misleading products being introduced in this market, and the difficulties and uncertainties facing functional foods, other than to suggest regulatory changes to allow for health claims on product labels as a panacea to 'free up' the market.

In this book we document what functional foods are; how they might work, and the markets, consumer trends and major players involved. We also look at the broader business environment that is emerging, particularly in the context of policies and regulations. We consider what we see as the clash between two nutrition revolutions

– that of the healthy-eating revolution, which started in earnest in the 1980s, and today's functional food revolution. In public policy circles functional foods and nutraceuticals are increasingly being described in terms of the 'medicalization' of food supply and are even satirized as '21st century quackery' (IACFO, 1999). In this larger context we argue that the future for functional foods is more open to both failure and success than many in the industry realize.

In short, we have attempted to bring a sense of order and perspective to what we argue is a 'revolution' which is transforming the food industry. This revolution aims to change the way we think about and consume food and drink. As we extend our analysis from nutrition policy, through regulation, to marketing activity, we argue that companies who want to succeed in the long term will need to adopt a radical new food-and-health-driven business model – a model which we outline in detail in Chapter 11.

Structure of the Book

Part 1 makes the case that functional foods are changing the face of food and health. In Chapter 1 we define and explain what functional foods are and why the subject has gained such immense importance for the food industry in a relatively short time. We set out the vision for functional foods and how this fits into developments in the international food economy. Finally, we outline the four main challenges presented by functional foods:

1 the development and marketing of functional foods;
2 acceptance by the consumer;
3 regulatory and policy issues; and
4 scientific and technological challenges.

In Chapter 2, we use a case study – of Benecol cholesterol-lowering products, invented and developed by the Finnish company Raisio – to illustrate how science, corporate ambition and nutrition marketing is pioneering this new relationship between food and health. Benecol is the product that is universally recognized within the food industry, and has come to symbolize the potential, the challenges and the meaning of functional foods.

Part 2 explains why public policy is the 'hot' topic in the field of functional foods. The functional foods revolution is characterized by the complexity of the issues it raises and the multidisciplinary skills needed to address them. We address what might be termed the public arena or policy context in which the functional foods revolution is unfolding. Included in the public arena are public health, nutrition policy, scientific and regulatory issues, and consumer advocacy.

Our approach to functional foods is unusual in that we ground our analysis in the totality of the food system and, as a result, emphasize the importance of the earlier healthy-eating nutrition revolution. This is detailed in Chapter 3, where we describe what we see as a clash of two nutrition revolutions. Starting around 25 to 30 years ago, countries and governments throughout the world developed and implemented healthy-eating nutrition policies based on a remarkable scientific consensus on the role of diet in a wide range of degenerative diseases, such as cardiovascular disease and certain cancers. The basic messages from this scientific consensus continue to be communicated extensively to consumers. We do not see functional food as a linear 'next stage' in food and health, but a fundamentally different type of food and health development.

From this point of view, we believe there is the potential for a fundamental conflict developing between the two nutrition revolutions. Such a conflict will have far-reaching implications for the future of functional food in the marketplace. It is a potential conflict that can only be resolved in the policy arena.

In Chapter 4 we turn to a key aspect of the functional foods revolution from the scientific and policy points of view – the risk of the use of such foods in relation to disease and health, and how the revolution is being used to communicate health benefits to consumers. We believe that in the development of functional foods, risk is often interpreted incorrectly from medical and other studies to apply to general populations. Nowhere is this confusion more apparent than with the distinction between relative and absolute risk. For functional food the notion of risk is central to:

- the construction of health claims;
- the development and use of biomarkers or endpoints;
- communication and marketing strategies about food and disease prevention that do not mislead the public; and
- the place of functional foods in public health policy.

Risk and disease and what these mean to ordinary people are central issues for functional foods. The whole premise of the functional foods revolution is that by consuming certain food components, the risk of disease is somehow reduced or avoided. But just what does risk prevention mean and how should functional foods be developed and marketed to optimize this?

Chapter 5 addresses one of the most complex topics in the development of functional foods – the use of health claims in food marketing and labelling. The functional foods revolution is blurring, or even redefining, the boundaries between food and medicine and how health is maintained or illness treated. This blurring is nowhere more apparent than in public policy or regulation, especially over the vexed issue of health claims – that is, what can and cannot be said about the health benefits of products. We outline the key issues and illustrate them with a detailed case study of public concerns about functional food and the development of the Joint Health Claims Initiative in the UK. Health claims are a theme that runs throughout this book.

Part 3 surveys the unfolding of the functional food revolution across the world. We explore the development of functional foods in three very distinct countries which are collectively leading the world in the functional food revolution – Japan, Finland and the US. While there are major differences between these countries, the common thread is the emergence of a new market for foods and beverages, which is being promoted on the basis of their health effects and benefits. These developments are often achieving the same goals – the creation of value-added markets, addressing of national diet-related disease and illness patterns and disease prevention. In all cases companies are effectively acting as providers of public health.

We start with Japan in Chapter 6 because it was here that the concept of functional foods was developed in 1984, and the Japanese market today is the most sophisticated in the world. Government has actively participated in this market solution to national health problems, an aspect which is often understated in analyses of the Japanese market. Uniquely, Japan has created a special regulatory category for functional foods and in doing so the term functional has been dropped in favour of Foods for Specified Health Uses (FOSHU). If a product meets the FOSHU conditions, a product-specific health claim is permitted.

In Chapter 7 we give an overview of the sea change in nutrition developments that has taken place in the US. These developments have been driven, in part, by two ground-breaking pieces of food regulation – the Nutrition Labeling and Education Act (NLEA) of 1990 and the Dietary Supplement Health and Education Act (DSHEA) of 1994. In the regulatory context of the US, functional food/nutraceutical developments are much more confrontational than in many other countries (thus continuing the pattern set by the 1980s healthy-eating revolution). This has been played out by lawyers and has usually pitted the US Food and Drug Administration against individual companies. The US provides an example of the importance of public policy in relation to the 'free market' development of functional foods. In this context, the US emerges as the surprising global pioneer of nutrition marketing and food policy.

In Chapter Eight we demonstrate that Finland is more than just Benecol, and describe its promotion as the 'Silicon Valley of functional foods', with a cluster of functional food businesses centred in the country. We argue that while Finland does have a cluster of functional food expertise and companies, it is underdeveloped and underperforming, and suggest that for it to capitalize on its unique functional food base, its cluster will need upgrading and a more robust strategy developed and implemented.

In Part 4 we delve deeper into the business and marketing realities of functional food. Chapter 9 considers in depth the business development and marketing of functional dairy products in Europe. The European dairy market has been the most proactive in developing and promoting functional foods. It has become the central battleground for some of the 'heavy hitters' of functional foods as companies fight it out for the minds and stomachs of the functional food consumer. The weapons of choice have been probiotic lactic acid bacteria for gut health and, what we call the 'battle of the little bottles', has become a most innovative and hotly contested area of market activity.

In Chapter 10 we detail the key strategic and marketing lessons from the international functional food revolution to date. We identify seven basic business strategies that are being adopted, with varying degrees of success, by companies trying to succeed in functional foods; these are:

1. the functional food make-over;
2. the fortified first-mover;
3. new product substitution;
4. the incremental new business creator;
5. the incremental old business developer;
6. whole category substitution; and
7. leveraging hidden nutritional assets.

We look, too, at the pricing of functional foods and their competitive position under the 'health proposition'.

In Part 5 we present a radical new business model – The Healthful Company™. It is the model for food companies who want to lead the functional foods revolution. It is our belief that companies who become 'truly healthful' can be at the forefront of the internationalization of food and health and be the pioneers of a new kind of food supply.

A Window on the Modern Food System

Functional food is more than a single trend, but part of a food product and nutrition revolution, the scope, scale and implications of which are only just becoming apparent. This book forms a benchmark against which future developments can be measured and assessed. Clearly, the topic of functional food cuts across many disciplines and issues and it is imperative to be aware of and understand its multidisciplinary context, whether it be from a business, policy or scientific perspective.

In covering these topics from an international perspective, together with extensive product, market and company case studies, this book is a first. It will be of interest to everyone involved in this emerging field – from product developers and marketers to policy makers, regulators, health and medical professionals, nutritionists and food scientists, in fact anyone who needs to understand how basic science is turned into commercial reality.

A Note on Methods and Sources of Information

Seeing the functional foods revolution unfold has been a personal voyage of discovery for us. In 1995 we founded the food industry journal, *New Nutrition Business*, to write about international marketing, policy and product trends in healthy eating and functional foods and beverages. Since then our work has evolved into a think-tank, the Centre for Food and Health Studies, which analyses, interprets and forecasts developments in food and health and advises corporations and consumer organizations.

In *New Nutrition Business* (see also www.new-nutrition.com), we have brought a unique perspective on healthy-eating developments, combining company, market, new ingredients and new product developments with consumer research, nutrition research and academic research in food policy and the global food economy. It is our readers – the executives and technologists of some of the world's largest food companies – and their activities which have inspired this book. Many, however, will not agree with our interpretations or analyses but others are actively embracing them. We have written this book because we want to have our say in what is becoming an increasingly complex business and policy environment. At a market level it is an environment in which a lot of money is being invested, lost or hangs in the balance. From a public health perspective, new functional food science and technology potentially offers immense benefit to large numbers of people. There is therefore an ethical imperative to communicate such information in a responsible and accessible form to as many people as possible.

In researching this book we have been confronted with a mass of material, often unrelated, contradictory or incomplete. Our purpose, in part, has been to reduce complexities in order to focus on the key issues. This has involved necessary omissions and narrowing of focus – for example, we look exclusively at foods and pay little attention to the many so-called functional soft drinks or to dietary supplements.

The subject of functional foods cuts across many disciplines, from different branches of science and medicine, nutrition, regulation, public health and policy to business development. We refer to the literature where appropriate, but we have not attempted to summarize vast areas of, for example, nutritional science or details of dietary components with 'health benefits' except to illustrate our key arguments. For those who wish to study further the *British Journal of Nutrition* presents an excellent overview (Bellisle et al, 1998).

In addition much of the material used in writing this book is what is often referred to as grey literature – including company reports, leaflets, booklets, conference proceedings, market and consultancy reports. We have also carried out numerous company interviews and visits. In this respect we have been helped by the many people in the food industry and academia who we have met and talked to over the past three years. We have also travelled extensively, throughout Europe, the US, Japan, Australia and New Zealand in pursuit of healthy eating and functional foods. We bring this personal, international insight to our analysis.

Michael Heasman
Helsinki, Finland
Michaelh@nutritiondigest.com

Julian Mellentin
London, UK
Julianm@nutritiondigest.com

1 May 2000

Part 1

A New Era in Food and Health?

Chapter 1

What are Functional Foods?

It is sometimes easy to forget that scientific knowledge about food and its role in health has its own history. This includes setbacks, apparent leaps forward, continuing challenges and breakthroughs, even fads and fashions, and of course, politics. It is an evolving, historically grounded body of knowledge, a true meeting of science and society. For example, it was in 1911 that Casimir Funk was telling the international community about a new discovery – the 'vital amine' – and its role in diet and health. Today vitamins, as they became known, are taken for granted as a part of our nutritional 'tool kit', even if their role in the diet continues to be the subject of intense research and scientific debate.

In the 1950s studies began in earnest on the relationship between nutrition and degenerative diseases, such as the possible connection between heart disease and dietary intakes of fat. This, at the time highly controversial, research agenda came to dominate nutrition policy in Western countries from the mid-1970s and continues today. This healthy-eating nutrition agenda – eating more complex carbohydrates while reducing intake of fats (particularly animal fats) and salt, and sugar – is just two decades old.

Recent years have seen the unfolding of an equally ground-breaking and new nutrition agenda, that of functional food science. Functional food science aims to maintain health, improve well-being and create the conditions for reducing the risk of disease. The targets, as always, are the so-called diseases of affluence – notably cardiovascular diseases and certain cancers, but also health concerns such as stiff joints, tiredness, physical performance, gut-related problems and allergies, for example.

Functional food science represents one of the more controversial areas of food and health because it suggests using food and the components of food in relation to the treatment or prevention of disease – the territory of drug development rather than food consumption. Functional foods have become a matter of public concern because of the food industry's rapid global drive to commercialize and market them, often before the supporting science is substantial or even, in the worst cases, has barely started.

Marketing Food, Nutrition and Health on an Unprecedented Scale

Functional foods should be seen as a concept that covers the marketing of food, nutrition and health and related products that is now taking place internationally on an unprecedented scale. In short, what is happening under the banner of functional foods is a revolution which gives a new and potentially ground-breaking 'spin' on the idea of good health through food. What is not always clear is whether this spin is in terms of individual health or directed towards the health of populations.

To begin to explain this spin we set out in this chapter what is meant by functional foods and why this has come to dominate the food industry and the food and health agenda. In Chapter 2, through a detailed market case study on the global introduction of a cholesterol-lowering margarine, we tell the story of how science, corporate ambition and nutrition marketing is pioneering this new relationship between food and health.

The food industry examples we use throughout this book are what we term serious attempts to develop scientifically-validated functional foods. Generally, these are designed for general sale and often have a public health 'edge', that is, they address recognized population-based diet-related problems. But, as in all food markets (and functional foods and beverages are no exception), there are some products of dubious merit and some which are downright misleading. These represent the darker side of functional food developments. We have chosen not to focus on these since they often detract from the bigger picture and the major issues. Our focus is mainly on the activities of some of the world's largest food corporations in the area of functional foods and where these might be leading.

A Chronology of Key Food, Nutrition and Health Events

While the link between food and health is nothing new, the concept of functional food only gained widespread recognition during the 1990s. The term itself was invented in Japan in the 1980s and the Japanese product 'Fibe Mini' – a soft drink containing dietary fibre – launched in Japan in 1988, is credited with being the world's first functional food. Seeing developments in Japan, some scientists and business consultants in the early 1990s started to actively promote the new trend of functional foods in the US and Europe.[1] But even in

Japan, the market for functional foods and beverages is still something relatively new. In a survey of Japanese product launches between 1988 and 1998 we identified 1721 functional products. However, 725 or 42 per cent of these were introduced in just two years (1997 and 1998); this rises to 946 or 55 per cent of all products surveyed if 1996 is included (see Chapter 6 on developments in Japan).

To set the scene, Box 1.1 depicts a chronology of landmark events in the development and history of functional foods.[2] Its purpose is to show the relative newness and evolving nature of functional food developments. While many of the events, companies and products mentioned in the box may be unfamiliar to some readers, they are explained in detail throughout the rest of this book. The chronology of food and health does not stop at 2000 – functional foods represent both an evolving science and food business activity. These together will change the way food is marketed in this new millennium.

A Definition of Functional Foods

Even though we prefer to see functional foods as a concept that captures the marketing of food, nutrition and health on an unprecedented scale by the food industry, the concept also extends to the scientific and nutritional role of food in relation to health. Despite this, and the high level of scientific activity geared towards unravelling the health-promoting secrets of food and food components, functional foods and nutraceuticals have defied consistent definitions among academics, scientists, business analysts and policy experts.

A simple definition, and one we like, has been put forward by the US Institute of Medicine. Functional food is defined as: 'any modified food or food ingredient that may provide a health benefit beyond the traditional nutrients it contains' (American Dietetic Association, 1995).[3] But behind such a simple definition lies a vast topic. It embraces global scientific discovery heralding what is being described as a 'new frontier in nutritional sciences' and regulatory and policy challenges that involve redrawing the boundaries between food and medicine (Heasman and Fimreite, 1998). Not least, food companies around the world are restructuring their operations and spending hundreds of millions of dollars to develop and market functional food and beverage products.[4] Functional foods symbolize the issues surrounding the wider topic of food and health and changes within the modern global food supply.

It is worth lingering over the definition of functional foods because this helps to illuminate why they are seen as so important to

Box 1.1 Chronology of Key Milestones in the Healthy-Eating and Functional Foods Revolutions

1950s onwards Researchers begin to investigate more closely the relationship between nutrition and degenerative diseases, such as the connection between heart disease and dietary intakes of fat. Landmark studies published by Ancel Keys et al establishing the link between heart disease, cholesterol and dietary fat, and Framingham study on heart disease (see Keys, 1953, 1980 and Dawber, 1980 on Framingham).

1955 Yakult Honsha Company incorporated in Japan.

1965 Concept of probiotic lactic acid bacteria introduced in paper by Lilly and Stillwell

1969 US White House Conference on Food, Nutrition and Health – one of the first high-profile meetings drawing public attention to the links between diet and the risk of chronic disease.

1976 Norwegian Government publishes world's first set of dietary guidelines.

1977 *Dietary Goals for the United States* (1977) published in February – one of the first documents to report several diet-disease hypotheses and to set quantitative target levels for reducing fat, saturated fat, and cholesterol in the American diet.

1980s Japanese invent term 'functional food' and step up research and funding.

By the end of the 1980s more than 100 expert and government healthy-eating reports, guidelines and goals had been published throughout the world. Food industry engages in frenzy of sugar-free, low-fat, and fat-reduced new product launches (this continues throughout 1990s).

1980 First edition of *Dietary Guidelines for Americans* published, subsequently updated every five years (*Nutrition and Your Health: Dietary Guidelines for Americans*, 1980)

1983 NACNE report published in UK giving Britons quantified dietary goals for first time (followed in 1984 by the COMA report on cardiovascular disease that saw government officially accept the diet and health link).

1983 Aspartame, the artificial sweetener – 200 times the sweetness of sugar – approved for use in European Union (EU)

1984 The Kellogg Company makes its landmark cancer prevention health claim for its All-Bran cereals in the US supported by the National Cancer Institute.

1988 Fibe Mini – a dietary fibre drink introduced in Japan by Otsuka Pharmaceutical – credited with being the world's first functional food.

1990 Concept of functional foods first introduced into Europe. Early 1990s sees spread of the concept in the food industry especially in US, but also in Europe

Finnish dairy company Valio launches Europe's first ever product range called Gefilus using a specific strain of scientifically-validated probiotic lactic acid bacteria Lactobacillus Goldin and Gorbach (LGG)

The NLEA becomes law and the US is the first country to introduce mandatory nutrition labelling. The Act also allows for the first time a number of generic health claims for food in relation to disease risk reduction that can be used on food labelling. The Act was implemented in May 1994.

1991 In September, the Japanese abandon the term 'functional food' and introduce the FOSHU system of self-regulation for food products which allows product-specific health claims.

1994 The DSHEA becomes law in the US. This allows structure–function claims for dietary supplements that do not need prior approval by the Food and Drug Administration (FDA).

First food science book specifically on functional foods published (Goldberg, 1994).

Cranberry juice and urinary tract infection study published (Avorn et al, 1994). Ocean Spray use the study to successfully promote cranberry juice in the US and sales increase by more than 150 per cent.

Functional Foods for Health Programme set up at the University of Illinois, US.

1995 Results from a study published in the *New England Journal of Medicine* show that Benecol – a margarine containing plant stanol esters – lowers (LDL) blood cholesterol by 14 per cent in Finnish men with elevated blood cholesterol.

In November, the Finnish company Raisio, launch 'Benecol' cholesterol-lowering margarine in Finland.

Japanese company Yakult Honsha build factory in The Netherlands and introduce the probiotic lactic acid bacteria fermented milk 'little bottle' concept to Europeans with 'Yakult' product launches in The Netherlands and Benelux countries.

Concept of prebiotics introduced for the first time in a paper by Gibson and Roberfroid published in *Nutrition Reviews.*

Four-year European Functional Food Science Programme, funded by the EU and managed by the International Life Sciences Institute (ILSI) begins.

Food Processing – a leading US trade press magazine – publishes industry survey which shows functional foods are the US food industry's top research and development priority.

Study published on lycopene – the substance that gives

tomatoes their red colour – and its potential to reduce the risk of prostate cancer in men. (For full discussion on lycopene see Gerster, 1997 and Clinton, 1998). Tomato Research Council set up to promote health benefits of tomato products.

1996 Australian Food Authority publishes its concept report on functional foods following a three-year consultation and review process on functional foods.

The FDA approves Olestra as a food additive – the 'fat free fat' developed by Procter and Gamble (P&G) (branded as Olean).

Yakult 'little bottle' probiotic drink introduced in the UK where it instantly becomes a success.

1997 In January, MD Foods axes its Gaio brand of cholesterol-lowering yogurt in the UK. It was first launched in June 1995 and one of the UK's first 'functional foods'.

In February the Quaker Oats Company petitions the FDA for a health claim for oats in relation to heart health, based on the cholesterol-lowering properties of soluble fibres. The petition becomes the first food-specific health claim to be approved by FDA.

The FDA extends oats health claim approval to cover pysllium husks and heart health following a petition from The Kellogg Company.

The Campbell Soup Company launches its ground-breaking Intelligent Quisine (IQ) range of clinically tested prepared meals for people with medical conditions such as heart disease and diabetes, spending a reported US$30–50 million in product development. The range fails and is withdrawn from market in June 1998.

Raisio signs worldwide licensing deal for Benecol with McNeil Consumer Healthcare, part of the US healthcare giant Johnson and Johnson.

In June, the UK Joint Health Claims Initiative is launched.

In November, Yakult announces plans to double European production capacity from 3.6 to 8.6 million bottles per week.

1998 In April, Unilever, the world's largest margarine producer, announces it has a cholesterol-lowering margarine based on plant sterols – the first direct competition to Benecol.

Herbal 'superstars' (such as St John's Wort, *Gingko biloba*, *echinacea* and kava kava) become main focus of product attention in US with soaring sales. Natural products boom in the US.

Australia starts a ground-breaking health claim pilot study using folic acid to test the power of health claims; more than 100 products carry the claim for folic acid in relation to neural tube defects in babies.

Heinz launch Lycopene Education Project in the US.

1999 UK Joint Health Claims Initiative publishes its draft guidelines for making health claims after two years of negotiations between consumer, food industry and enforcement authority representatives.

In May, both Raisio's Benecol and Unilever's Take Control (called Pro.Activ in Europe) cholesterol-lowering products introduced in the US.

More than 3000 dietary supplement products using structure–function claims in US.

Novartis launches a complete range of functional foods called Aviva for bone, heart and digestive health in the UK and Switzerland.

In July, the FDA approves a health claim for whole grains in relation to heart health and certain cancers following a notification by General Mills using the FDA Modernization Act of 1997 – the first successful health claim petition using this legislation.

In October, the FDA approves a health claim for soy protein and heart health following a petition submitted by Protein Technologies International (part of the DuPont group) based on a scientific dossier that shows the ability of soy protein to lower cholesterol.

Scientists suggest that the recommended daily allowance (RDA) for vitamin C in the US should be increased substantially, based on new science showing the health-promoting benefits of antioxidant vitamins.

At the end of 1999 the Finnish dairy company Valio has secured licences for the use of its probiotic LGG in products in 26 countries.

Danone starts test-marketing 'Actimel' 'little bottle' probiotic in the US; Yakult rumoured to be test-marketing their 'little bottle' in California.

At the end of the year there were 159 FOSHU-approved products in Japan in FOSHU market worth around US$1 billion. More than 80 per cent of FOSHU-approved products are for 'gut health' (174 FOSHU approvals up to February 2000).

2000 (to May) Raisio and McNeil announce radical changes in strategy to prop up ailing Benecol range.

Unilever able to launch their cholesterol-lowering margarine in Europe (in August) following a 17-month delay resulting from the EU novel foods process.

The Kellogg Company, in a radical labelling move, announce they will make a structure–function claim in the US on a selection of their adult cereals for folic acid and cardiovascular health.

In July Novartis withdrew Aviva range in UK due to poor sales. Continues on sale in Austria and Switzerland.

A proliferation of products making heart health and cholesterol-lowering claims in the US and UK.

Continuing expansion of the 'little bottle' probiotic concept.

Source: New Nutrition Business (various years)

the future commercial health of the food industry as well as, potentially, the health and well-being of the general population. The first book bringing together the science behind the different areas of functional foods was published as recently as 1994 (Goldberg, 1994). Functional foods were defined here as: 'any food that has a positive impact on an individual's health, physical performance or state of mind in addition to its nutritive values'.

In a special supplement of the *British Journal of Nutrition* on Functional Food Science in Europe (Bellisle et al, 1998) a food is said to be functional if it 'contains a food component (whether nutrient or not) which affects one or more targeted functions in the body in a positive way'. It could also include foods in which a potentially harmful component has been removed by technological means. In other words, functional foods offer the potential for far-reaching health benefits. Implicit in both definitions is that this is in the form of some kind of processed food or drink, usually with something added for health effects.

Within the US in particular, the concept of functional foods fell upon fertile ground. For example, the term 'designer foods' was coined in 1989 by Dr Herbert Pierson, then director of the National Cancer Institute's US$20 million Designer Foods Programme, to describe foods which naturally contain or are enriched with non-nutritive, biologically-active chemical components of plants (phytochemicals) that are effective in reducing cancer risk. Also in 1989 in the US, Dr Stephen DeFelice, chairman of the Foundation for Innovation in Medicine, invented the term nutraceutical, to refer to any substance that may be considered a food or part of a food and provides medical or health benefits, including the prevention and treatment of disease. More formally, the Nutraceuticals Institute has defined nutraceuticals as 'naturally derived bioactive compounds that are found in foods, dietary supplements and herbal products, and have health promoting, disease preventing, or medicinal properties' (Nutraceuticals Institute, 1998).

This discussion concerning the actual words and definitions used to describe these new food and health developments is characteristic of the functional foods/nutraceuticals debate, especially as companies and policy makers struggle to enable health claims to be made for products. Often nutraceutical is used to describe the bioactive ingredient that can deliver a health benefit and which could be used in a capsule, pill, powder or foods and beverages – thus you get a functional food. However, the terms functional food and nutraceutical are often used to mean the same thing, with a preference in the food industry for the term functional food, while nutraceuticals are increasingly being associated with dietary supplements (that is, pills and capsules). You are free to choose your favourite term! The important point, is that currently there is no legal definition for functional food, beverage or nutraceutical in Europe or the US.

Foods for the Healthy or Foods for the Sick?

It remains uncertain whether functional foods, by these definitions, are seen as foods for healthy individuals, or those who are ill or have a medical condition, or for both. This important distinction needs to be made, since the marketing of functional foods is taking two distinct routes. The first is the development of products for people who are ill with a specific and medically recognized health-related condition, for example, elevated levels of cholesterol. In extreme cases this would also include foods for people with more serious illnesses or disease, such as cancers or other conditions that require special dietary needs. In other words functional foods do aim to treat, cure, prevent or mitigate specific illnesses or disease. The second route is of products marketed for healthy people, which may possibly prevent disease but also enhance the health status of already healthy individuals, for example, improving the 'healthy' environment of the human gut. These two very distinct roles and targets for functional foods are often confused as one and the same thing.

Functional Food – a Convenient Tag

More recently, the concept of functional food has in fact become more blurred and moved beyond the definitions discussed above. A whole range of traditional foodstuffs – from oats and tomatoes to soy and whole grains – are now being marketed on the basis of their health-promoting properties, often established through years of scientific research on their specific dietary components. It is fair to question whether these really are functional foods, since they have not been modified or had anything added to them, and they have been around for centuries. But it is the relatively new scientific validation of the health-promoting components of these traditional foodstuffs, the aggressive marketing of this new nutrition information, together with strong health claims being made concerning disease risk reduction (both approved and non-approved), that places them under the collective umbrella of functional foods.

We have lingered over defining a functional food because there is confusion and a divergence of opinions over this term. We have tried to demonstrate that functional food is a powerful conceptual tool to describe a fundamental new direction in food, nutrition and health science, as well as food marketing. The term is, therefore, a convenient tag, and is used throughout this book. Unless otherwise stated, we focus on foods and beverages rather than pills, capsules or dietary supplements. It is also important to note that functional food does not describe a distinct food category as used in conventional market analysis, for example confectionery or snack foods, and therefore functional foods do not sit on a specific shelf in the supermarket. To summarize, the functional food concept describes a fast-growing range of food industry developments that cut across traditional food and beverage categories. Original definitions of functional foods see them as part of the normal diet or food pattern and recognized as being beneficial for well-being and health. Today, functional foods can be seen as products with something 'added' ('extrinsic' functional food) or foods and beverages promoted because of their 'intrinsic' health benefits (for example 'whole grains').

Raising the Profile of Functional Foods

In raising awareness about the concept of functional foods, in both Europe and the US, the International Life Sciences Institute (ILSI) has been instrumental in developing a science-based approach to functional foods. For example, it has organized international forums such as the First International Conference on East–West Perspectives on Functional Foods in 1995.

But there have been other pioneers – from individuals to food companies and academia. Notable within the US has been the Functional Foods for Health Program (FHP) based at the University of Illinois at Chicago and Urbana-Champaign under the leadership of Dr Clare Hasler and (until his retirement) Professor Norman Farnsworth, first set up in 1994. Dr Hasler, as executive director, has created national recognition for the FHP and gained support for the programme from a virtual 'who's who' of industrial affiliates among American companies with an interest in functional foods. Box 1.2 lists industrial affiliates to the FHP.

It is interesting to note the mission statement of the FHP, which reads:

> *'Functional Foods for Health ... is dedicated to the improvement of human health and the reduction of health care costs through research and education related to the identification of food components and the development of food products which have disease-*

Box 1.2 Industrial Affiliates of the Functional Foods For Health Program, University of Illinois, US, July 1999

Abbott Labs; Almond Board of California; Archer Daniels Midland Company; California Prune Board; Cargill; ConAgra, Inc; General Mills; Hershey Foods Corporation; Kraft Foods, Inc; McNeil Consumer Health Care; McNeil Speciality Products Company; Mead Johnson Nutritional Group; Monsanto; Nabisco, Inc; National Honey Board; Nestlé USA, Inc; Novartis Consumer Health; Ocean Spray; Pharmavite Corporation; Protein Technologies International; Pulsar/Seminis Vegetable Seed Co; Reliv International, Inc; Ross Products Division; SunStar, Inc; Tropicana North America; Warner Lambert; Welch's;

Source: FHP, 1999

> *preventative and health-promoting benefits'* (Functional Foods for Health Program, 1999).

Another university-based initiative in the US is The Nutraceuticals Institute, an alliance of Rutgers, the State University of New Jersey and Saint Joseph's University, Philadelphia, set up under the leadership of founding director Dr Nancy Childs and executive director Professor Paul LaChance.

In Europe, the functional food 'tower of strength' has been Professor Marcel Roberfroid, based at the Universite Catholique de Louvain, Brussels, Belgium, and past president of ILSI. Professor Roberfroid was scientific coordinator for the European Commission (EC) programme of work on Functional Food Science in Europe; which began in 1995 and is being managed by ILSI Europe. But the pursuit of functional food science is not just an abstract scientific or university project. For example, included in the objectives of the EU functional food science programme is the following:

> *'Special attention will also be given to ensure the transfer of knowledge towards small and medium scale enterprises.* It is finally expected that the project will lead to increased competitiveness of the European food sector in the context of a growing world market *and to a greater understanding of the role and potential of functional foods for improving people's diets and health'* (ILSI undated, emphasis added).

In addition, in Europe, all the major food research institutes, such as TNO in The Netherlands, VTT Biotechnology in Finland and the Institute for Food Research in the UK, have dedicated research programmes or projects on functional foods as do most major food and ingredient companies.

The Vision for Functional Foods

Why is there so much interest in functional food, especially when no one can agree on a definition? There are three major factors driving the functional food revolution:

1. A truly ambitious health vision for the developed and developing world
2. Food companies have bought in to the market potential of a new type of consumer for health
3. Shareholder imperatives drive corporate ambition in functional foods. In highly competitive food markets with tight margins and slow-growing food sales – but shareholder demands for profit growth – functional foods are seen as a way to achieve added-value growth and profitability

A truly ambitious health vision

For the developed world the 'vision' of functional foods, if ever realized would mean a fundamental change in diet. Functional foods are about manipulating and constructing foods and diets not to just maintain well-being or a balanced diet, but to actively participate in shaping health status. The fundamental difference between the functional foods revolution and current nutrition practice is that functional food aims to provide certain dietary components which may be difficult to obtain through a normal modern diet in quantities which can positively affect health.

An example to illustrate this point is folic acid/folates[5] and the prevention of neural tube defects in babies. Dietary surveys have shown that many women who become, or might become, pregnant do not get the quantities of folate through normal dietary practice that have been scientifically shown to help reduce the risk of babies being born with spina bifida (it is recommended women who are pregnant or may become pregnant take a daily intake of folate of 400 mg). Although advocates of functional foods are not suggesting the general fortification of foods, folate is cited as an example of the principle whereby a dietary component (in this case folate) with a proven health benefit is difficult to obtain in sufficient quantities through the normal diet to achieve the health benefit status. In the case of folate it has been incorporated into general food supply through food fortification. Functional foods, through consumer choice rather than general fortification, seek to deliver a recognized health benefit that is difficult to achieve through normal dietary practice.

Folate illustrates the power of functional food science to change our understanding of food in relation to health. There is now compelling evidence for an association between elevated blood levels of homocysteine and the risk of cardiovascular disease. In addition there is a relationship between intakes of folate and the vitamins B6 and B12 in reducing homocysteine levels and therefore the potential, through homocysteine-lowering dietary therapy, to reduce the risk of cardiovascular disease. The causality of the relationship between elevated homocysteine plasma levels and cardiovascular disease remains controversial and well-designed dietary intervention trials to establish the effectiveness of folate and vitamin B6 and B12 still need to be undertaken (for a full discussion see Hornstra et al, 1998; Meleady and Graham, 1999).

However, in April 2000, The Kellogg Company, a long-time advocate of the association between folic acid and cardiovascular disease, announced it was making a 'structure-function' claim linking folic acid, B6 and B12 and cardiovascular health on the packaging of a number of its adult breakfast cereals in the US (see Chapter 7).

The functional food revolution is therefore, in one sense, about constructing an 'unbalanced' diet skewed towards optimum health. In the language of food marketing, the focus is now on supplementing the diet with something 'good' – about adding benefits – rather than removing something 'bad', such as fat and sugar. In short, functional foods imply that there are now 'good' foods in the diet, although this view is strongly challenged by many in the food business (and, strangely, there are still no 'bad' foods, only poor diets).

Put simply, the scientific challenge is all about identifying the individual components of plants and other foods that can prevent disease and illness and also enhance and prolong healthy and active life. It means developing diet-based disease prevention strategies and formulating foods tailored to meet specific health needs. A key goal of functional food science is to understand and target particular 'biomarkers' or endpoints for disease and illness and the food components that act upon these (see Chapter 4). Stripping away the scientific jargon, this could mean, for example, delaying the ageing process;

stopping certain cancers from forming; reducing the risk of diseases such as cardiovascular disease; improving the immune system; maintaining and improving physical and mental functioning; and many more such dietary interventions to 'optimize nutrition' (Milner, 1999).

The functional foods revolution doesn't just stop at the developed world. In developing countries it could mean delivering vaccines through food products such as bananas, or growing rice with high concentrations of vitamin A to help prevent blindness, and other such nutritional interventions (although this aspect of functional food is not addressed in this book).

The functional foods revolution is truly a global, scientifically-grounded, high-technology, but marketing-led vision of future manufactured food supply.

A Fundamental Shift in Nutrition Thinking?

The 'holy grail' of science and business is to find the actual component in a food that influences human health and the relevant human biomarker it acts upon. Taken to its logical conclusion this is a fundamental shift in the use and application of nutrition science. Another recent example serves to demonstrate this shift and what functional food science is all about.

In April 1999 US government researchers, publishing in the *Journal of the American Medical Association* (JAMA) recommended that the daily allowance for vitamin C should be doubled or tripled in the light of increasing evidence of its anti-cancer potential (Levine et al, 1999). The established RDA for vitamin C is 60 mg per day and is still based on preventing scurvy, but the researchers from the National Institute of Health say that since 1989 extensive new biochemical, molecular, epidemiologic and clinical data have become available. The authors point out that diets with 200 mg or more of vitamin C from fruits and vegetables are associated with lower cancer risks, especially for cancers of the oral cavity, oesophagus, stomach, colon and lung.

The researchers, therefore, proposed a new RDA of 120 mg/d and an 'adequate' intake of 200 mg/d. They suggest an upper limit of 1g. Vitamin C is now no longer a 'deficiency' problem in the diet that gives rise to scurvy but can also promote health and prevent disease from developing when consumed in sufficient quantities. Importantly, they also propose that the higher vitamin C intakes should come from eating at least five portions of fruit and vegetables daily, an amount they say appears to be protective. They speculate that fruit and vegetable intake may be associated with lower cancer risk, not because of vitamin C alone but 'perhaps because of interactions between ascorbate and bioactive compounds in these foods, or because of compounds independent of vitamin C, or because of characteristics of people who eat fruits and vegetables'.

Functional food product development means using science to develop new products that deliver the benefits of antioxidant vitamins, like vitamin C. The reasoning is that certain natural dietary components, such as vitamin C or the example of folic acid cited earlier, have a scientifically validated health potential beyond basic nutrition. Product development is taking place using a whole range of dietary components. Popular in recent years have been calcium, plant sterols, the antioxidant vitamins, and a host of other ingredients, some of which we describe later. It is argued that the quantities of such dietary components needed to deliver these newly established health benefits are often difficult to obtain in the modern or 'normal' diet as generally consumed. Importantly, in most discussions of functional food these are seen as foods for everyday use, and as part of general food supply.

The Scope and Challenge of Functional Foods

The scope and scale of the potential for functional food is staggering and the full spectrum of human disease and illness is under investigation. Table 1.1 summarizes areas of scientific enquiry underway in Europe in relation to functional food, while Table 1.2 lists some of the actual foods being studied for their health effects. Functional food could be targeted at virtually all diet-related disease and ill heath and the diets of healthy individuals. It already includes most of the modern food supply from dairy products, grain, cereals, fruit, vegetables and fish and has recently been extended to meat. Functional food science and product development has the potential to reshape our food supply.

The New Consumer Health Trend

One reason why we believe the functional foods revolution is unstoppable is that the world's food industry has bought in to what it sees as a new consumer relationship with health, and this includes attitudes to food. Allied to this, in developed countries, food companies believe the marked demographic changes that will appear over the next 30 years, in particular an ageing population, will reshape food markets.

Table 1.1 *Main Areas of Scientific Investigation in Europe in Relation to Functional Food*

Human growth and development (especially among infants and young children and during pregnancy)
Obesity
Diabetes
Cancers
Malnutrition (for example, among the elderly)
Exercise and diet
The role of antioxidants in disease prevention (especially vitamins A, C and E, the flavonoids and carotenoids)
Cardiovascular disease
Gastro-intestinal function and physiology
Behaviour and psychological functions

Source: Adapted from Bellisle et al, 1998

A host of age-related health statistics are cited to justify the functional food revolution. The 'baby boom' generation (defined loosely as people aged 35 to 54) is in particular seen as the target for new functional food products. In the US, for instance, 78 million 'boomers' are heading for what they are constantly reminded are their most disease-prone years. At current rates the number of Americans with osteoporosis or low-bone density will hit 41 million by 2015, up from 28 million in 1999. Heart attacks, which struck 1.1 million Americans in 1999, will strike 1.5 million by 2025. Stroke incidence, about 400,000 in 1999, will top an estimated one million by 2050 (Cowley, 2000). Functional foods are presented as the 'nutritional solutions' to such frightening statistics.

What is more, companies are studying international consumer research which they believe describes a new type of consumer for health-related products. For example, more and more people are becoming interested in preventative self-medication rather than therapeutic health strategies; they behave as being empowered and nutritionally aware and are seeking personal control of their own health and well-being. In countries such as the US, the staggering personal financial cost of healthcare is also seen as a strong motivating factor.

Health is the 'Heavy' Trend

Health is the 'heavy' trend in modern food eating. This was one of the main themes of a presentation on changing habits of the consumer

Table 1.2 ***Examples of Functional Food Components and their Potential Benefits***

Class/ Component	*Source* [a]	*Potential Benefit*
Carotenoids		
Alpha-carotene	Carrots	Neutralizes free radicals which may cause damage to cells
Beta-carotene	Various fruits, vegetables	Neutralizes free radicals
Lutein	Green vegetables	Contributes to maintenance of healthy vision
Lycopene	Tomatoes and tomato products (ketchup, sauces, etc)	May reduce risk of prostate cancer
Zeaxanthin	Eggs, citrus, corn	Contributes to maintenance of healthy vision
Collagen Hydrolysate		
Collagen Hydrolysate	Gelatine	May help improve some symptoms associated with osteoarthritis
Dietary Fibre		
Insoluble fibre	Wheat bran	May reduce risk of breast and/or colon cancer
Beta glucan [b]	Oats	Reduces risk of cardiovascular disease
Soluble fibre [b]	Psyllium	Reduces risk of cardiovascular disease
Whole grains [b]	Cereal grains	Reduces risk of cardiovascular disease
Fatty Acids		
Omega–3 fatty acids – DHA/EPA	Tuna, fish and marine oils	May reduce risk of cardiovascular disease and improve mental, visual functions
Conjugated linoleic acid – CLA	Cheese, meat products	May improve body composition, may decrease risk of certain cancers
Flavonoids		
Anthocyanidins	Fruits	Neutralizes free radicals, may reduce risk of cancer
Catechins	Tea	Neutralizes free radicals, may reduce risk of cancer
Flavanones	Citrus	Neutralizes free radicals, may reduce risk of cancer
Flavones	Fruits, vegetables	Neutralizes free radicals, may reduce risk of cancer

Class/ Component	*Source* [a]	*Potential Benefit*
Glucosinolates, Indoles, Isothiocyanates		
Sulphoraphane	Cruciferous vegetables (broccoli, kale), horseradish	Neutralizes free radicals, may reduce risk of cancer
Phenols		
Caffeic acid Ferulic acid	Fruits, vegetables, citrus	Antioxidant-like activities; may reduce risk of degenerative diseases; heart disease, eye disease
Plant Sterols[b]		
Stanol ester	Corn, soy, wheat, wood oils	Lowers blood cholesterol levels by inhibiting cholesterol absorption
Prebiotics/Probiotics		
Fructo-oligo-saccharides – FOS	Jerusalem artichokes, shallots, onion powder	May improve gastrointestinal health
Lactobacillus	Yogurt, other dairy products	May improve gastrointestinal health
Saponins		
Saponins	Soy beans, soy foods, soy protein-containing foods	May lower LDL cholesterol; contains anti-cancer enzymes
Soy protein		
Soy protein [b]	Soy beans and soy-based foods	25 g/d may reduce risk of heart disease
Phytoestrogens		
Isoflavones – Daidzein, Genistein	Soy beans and soy-based foods	May reduce menopausal symptoms, such as hot flushes
Lignans	Flax, rye, vegetables	May protect against heart disease and some cancers; lowers LDL cholesterol, total cholesterol and triglycerides
Sulphides/Thiols		
Diallyl sulphide	Onions, garlic, olives, leeks, scallions	Lowers LDL cholesterol, maintains healthy immune system
Ally methyl trisulphide, Dithiolthiones	Cruciferous vegetables	Lowers LDL cholesterol, maintains healthy immune system
Tannins		
Proantho-cyanidins	Cranberries, cranberry products, cocoa, chocolate	May improve urinary tract health and reduce risk of cardiovascular disease

Notes: a Examples only – not all-inclusive
b FDA-approved health claim established for component
Source: International Food Information Council, Washington, DC, 2000

given by Larry Hasson, managing director of the French-based International Research Institute on Social Change (RISC) given at the European Food 2000 Seminar in Helsinki, Finland in September 1999.

In an analysis based on RISC's Anticipating Change in Europe Programme – an annual questionnaire of some 140 socio-cultural questions administered to a representative population sample of 2500 in the five main European markets – Hasson described what he calls the different 'logics' of the European food consumer.

He explained that health today is 'an expression of modernity and food innovation' and represents a 'holistic view of the body, mental as well as physical'. The consequence for products, he said, was that 'a bad image on health can lead to the rapid disappearance of a product from the market'.

The major trends for food which RISC identified are:

- Pleasure
- Quality (especially for people aged over 50)
- Innovation
- Compulsion (associated with the young and 'masculine logic')
- Continuity (the number of product 'supporters' increases with age, especially after 50)
- Health

Hasson broke down the 'health' trend further into five main areas of consumer concern:

- Expertise (people want diet and nutrition know-how)
- Morals (for example, fat and sugar phobias)
- Caring (people want to be proactive, for themselves and others)
- Compensation (the 'plus' products, for example functional and enriched foods)
- Restriction (the 'less' products, for example, less fat)

It is interesting to note how the 'health' trend is expressed in different European countries. For example, the French and Italians are Europe's health and nutrition 'experts' and are more attentive to nutritional information. On the other hand, the British and Germans are least worried about their health, but buy most 'medical' type products!

In addition, numerous market studies and consumer surveys in the US, such as those carried out by the Food Marketing Institute or the American Dietetic Association's bi-annual nutrition trends survey, show that American consumers are concerned about food, nutrition and health and are prepared to change their diets accordingly. What is not so clear from such surveys, however, is whether consumers are as yet buying into the functional food concept. But it is the well-documented trends about consumers' concern over diet and nutrition that food companies are seeking to exploit through functional foods.

Shareholder Imperatives Drive Corporate Ambition in Functional Food

While functional foods are changing the face of the nutrition map, we should not lose sight of why the food industry is particularly interested in such developments. Functional foods, or more correctly food and health issues, have to be seen in the context of a global food industry faced with falling profit margins in traditional product markets, and an industry driven by the new challenges and economics of shareholder imperatives. This usually means that institutional shareholders want to see double-digit growth earnings on their investments. With the explosion in value of the technology, telecommunications and similar sectors and the excitement of the 'new economy', stock markets are regarding 'old economy' companies, like many food companies, as poor investments and as lacklustre performers (even though most food companies actually make profits, have customers and make things that people want to buy). The brutal financial truth is that the shares of many food companies have underperformed in recent years, increasing the pressure on food companies to come up with something new and financially exciting for investors.

Food companies have responded to demands for profit growth with massive restructuring, extensive cost cutting and a frenzy of mergers and acquisitions to grow market share and consolidate product categories. Companies have also applied immense pressure to derive 'value-added' benefits from products and categories in what are often saturated, static or declining markets (people can only eat so much). It is worth digressing to consider this aspect of the modern food system in more detail and how it relates to food and health.

What Characterizes the Current Global Food System?

The global dynamics of the modern food system provide the backdrop to both the healthy eating and functional foods revolutions described in Chapter 3.

The 20th century witnessed one of the most significant food revolutions since settled agriculture began around 10,000 years ago.

Beginning in the UK and US agricultural heartlands, radical changes in how food was grown, processed, distributed, sold, cooked and even eaten were experimented with, applied and marketed. The century began already well-endowed with processing technology such as the giant roller mills used in grain milling, and with industrialized baking machinery enabling, for example, the mass production of biscuits by the million.

But these changes were nothing compared to what was to follow. Farming was industrialized, with giant machinery replacing human labour and the application of what social scientists call Fordist[6] thinking to both plant and animal production. Extraordinary creativity was expended on trying to reduce nature's unpredictability (Goodman and Redclift, 1991). Agrochemicals replaced the hoe. Feedlots replaced grazing. Monoculture replaced smallholdings. These changes are all well documented, and owe much to the spread of fossil-fuel culture, in particular the use of oil to drive machines. Less commented upon, until more recently, have been the changes further down the supply chain.

If the first half of the 20th century was marked by industrialization of both agriculture and processing, the second half will surely be enshrined as the period of retailing and culinary industrialization. New ways of packaging, distributing, selling, trading and cooking food were developed, all to entice the consumer to consume. These power shifts in the food economy contributed to the contemporary conflicts within the food system between the productionist and consumerist sectors – for example, agribusiness versus consumer business; primary producers versus traders; food processors versus food retailers; and even production interests versus public health goals.

In addition, it is sometimes easy to forget that the past 20 years have seen unprecedented changes in the availability of, and access to food, especially in Western Europe and the US. One major change has been the year-round variety of foods now available from all corners of the world and the virtual elimination in the seasonality of fruits and vegetables.

However, reduced to its bare essentials the global food economy is built upon a relatively simple base. Most of us eat food from a core group of about 100 basic food items, which account for 75 per cent of our total food intake (Sims, 1998). As Sims writes:

> *'We may think we buy foods for their sensory appearance, freshness, safety, nutritional quality, healthfulness, convenience, or price. Those factors undoubtedly affect our food purchases, but we are still constrained in our choices by the routines, habits and*

associations that have surrounded our interactions with food throughout our lifetimes.'

From this base we have created a food economy of immense size and economic power. Nestle (1999) summarizes some of the pertinent food facts and figures for the US. For example, food and beverage sales earned US$890 billion in 1996, of which nearly half was spent on meals and drinks consumed outside the home. But only 20 per cent of food expenditure (the 'farm value') went to food producers, the remaining 80 per cent constituting added value in the form of labour, packaging, transportation, advertising and profit. In 1996, the US marketplace included 240,000 packaged goods from US manufacturers alone. New product introductions in 1996 totalled 13,600, 75 per cent of these were candies, condiments, breakfast cereals, beverages, bakery products and dairy products. These and other foods are advertised through the electronic and print media at an annual cost of more than US$11 billion with another US$22 billion being spent on coupons, games, incentives, trade shows and discounts. In 1997, advertising for a typical candy bar required expenditure of US$25–70 million, and for the burger chain McDonald's, nearly US$1 billion.

An important consequence of such economic activity is that the American food supply provided 3800 kcal/day for every man, woman and child in the country, an increase of 500 kcal/day since 1970. This level is nearly twice the amount needed to meet the energy requirements of most women, one-third more than that needed by most men, and far higher than that needed by babies and young children. 'These figures alone describe a fiercely competitive but slow-growing food marketplace, one in which food companies compete to sell more of more profitable foods' (Nestlé, 1999).

The key features of the modern food economy include:

- A rapid concentration in all sectors, both through organic growth and mergers and acquisition within and across national borders;
- A fragmentation of markets;
- Comparatively rapid, commercially driven changes in diet and taste;
- A ceaseless pursuit of quality control via supply chain management intensification both on and off the land;
- Widespread application of science and technologies to food production, through farm to factory to distribution;
- Transformation of foods and food processes across sectors – not just the nature of farming and storage have been transformed but even cooking;

- The growth in size and influence of the distributors and retailers within the food system, representing a transition from producer to retail power;
- An ideological tension over the state's role and responsibilities both in law enforcement and in public education;
- Unmanageability in the consumer body politic, with a growth of consumerism threatening predictability for dominant forces within the food system;
- Promotion of brands as a response to consumer desire for certainty;
- Ever more sophisticated use of the conscious industries (marketing, advertising, product placement, sponsorship) to increase brand value;
- New inequalities within and between countries creating modern forms of food poverty even in rich countries;
- Centralization of decision making nationally, regionally and internationally, with tensions between all levels; and
- A pivotal battle for world markets between the EU and the US.

Although globalization is a much-used word even in the context of the food economy, it should be stressed that much food eaten around the world is still nationally or regionally produced and trade is limited to relatively few countries. The concern here is about the tendency to globalize, and this is a major driver for functional foods and ingredients with a constant global quest to find the next 'hit' product or ingredient. We illustrate these globalizing tendencies in Chapter 2 through our case study on Benecol and in Chapters 9 and 10, which describe business strategies. In short, the food industry is faced with three major choices – globalize, vertically integrate, or move into new sectors. Food, nutrition and health, but especially functional food, offer one of the few opportunities for companies to globalize and move into new sectors, as well as 'add value' to traditional foodstuffs.

A further pressure on food processors is the power of the supermarket. Today we live in a hypermarket economy, and with giant retailers such as the US-based Wal-Mart, the Dutch retailer Royal Ahold and the French company Carrefour all expanding internationally, the 'mega-market' economy looms. Observers have noted that in this transition, consumers find themselves partly acting as service providers for the company. For example, as Ritzer (1992) observes, this may mean literally walking miles down vast aisles to find a few goods.

Of developed countries, Australia has the highest supermarket concentration. Three companies – Woolworth, Coles Myers and Franklin – have an estimated 80 per cent of national sales. In the UK, four companies (Tesco, Sainsbury's, Asda and Safeway) have around 44 per cent (IGD, 2000). Food retailing in most rich countries is characterized by major changes in technology, location and legal frameworks. In Germany, for instance, the laws on weekend opening hours were changed to suit large companies, but not family-owned businesses. In Japan, a similar epic struggle has been going on to change regulations on opening hours.

Such changes have not gone unnoticed. In Australia, in 1999, a Joint House Select Committee Inquiry into supermarket power was set up. In April 1999, in the UK, the Office of Fair Trading referred allegations of price fixing by supermarkets to the new Competition Commission, which found on a number of counts that UK supermarkets operated against 'the public interest'. The dynamics of the market and hypermarket economies are very different. In the hypermarket economy, supply chain management is the key.

Retailers are sovereign because they mediate between production and consumption – any functional food will have to meet retailer expectations. It is interesting to note that the UK's leading food retailer, Tesco, is said to have been reluctant to stock Benecol at first, until consumers asking for it prompted them to change their mind. Another functional product, Yakult (see Chapter 9), saw its marketing strategy turned on its head when the major UK retailers suddenly demanded national distribution, rather than the careful regional marketing campaigns the company had wanted.

The Geographical Building Blocks of the Current Food System

The shape of today's food system has been heavily affected by policy decisions made east and west of the Atlantic in the aftermath of the Second World War. Both today's food 'superblocs', the US and EU, have policies of farm support designed to avoid the suffering of the war period. Although different in format, both adopted 'productionist' frameworks. The state encouraged the development of agribusiness and infrastructural industries whose purpose was to make labour and land yield more – to be more capital efficient. Today's ecological and consumer confrontations with technical and scientific developments such as biotechnology, agrochemicals or factory-farmed foods have their roots in reactions to the policy choices made in response to the recession of the 1930s and the recovery from the war.

Supporters of the agricultural revolution rightly point out that, despite massive population growth, supply has kept pace with demand. The Food and Agriculture Organization (FAO) estimates that there is more than enough food, by calories, to feed the world's population adequately. Yet in the period from 1994 to 1996, 19 per cent of the population in the developing world were undernourished. Even when proportions of undernourished people declined, in absolute terms numbers often grew. Thus, although there is a theoretical food security, it is not achieved in practice. Distribution, land ownership and control over food have distorted the possibilities.

Historically, there was also the divorce of health from agriculture, particularly after the Second World War. More than half a century ago the global marriage between health and agriculture almost took place with the setting up of the FAO. Sir John Boyd Orr, a Scottish nutrition researcher, had been pioneering the symbolic marriage of health and agriculture for many years and by 1945 the idea of joining health and agricultural policy goals had already been considered by scientists from all over the world (Bengoa, 1997). In 1943, during the War, the Conference on Food and Agriculture in the US highlighted the need to have the agriculture and health sectors collaborate to confront the world's growing nutrition and agriculture problems. Another conference goal was to help lift US farmers from the aftermath of the 1930s dustbowl and recession. Boyd Orr's vision was global: to marry need with farming surpluses (Boon, 1997). It was a classically top-down perspective, which today would sit a little uneasily with agriculture and food policy talk of community participation and people-led visioning.

Nevertheless, there is much to learn from it. Two reasons why the marriage of agricultural and health policies failed to take place have been suggested (Bengoa, 1997). The first is the lack of professionals to study, analyse and execute strategies and evaluate national nutrition problems. Secondly, and just as telling, the marriage did not occur right away at a national level owing to lack of interest in nutrition displayed by the agricultural ministries. It might be argued that much public health and nutrition policy as well as food industry activity over the past 20 years is an attempt to restore this historically missed opportunity.

The functional food revolution provides another opportunity to reopen discussions on the marriage of health and agriculture, especially through the development of fruit, vegetables and basic foodstuffs with enhanced health-promoting benefits. Food and health concerns offer one way to open up new choices for agriculture as well as the food industry. For example, one of the drivers behind setting up the US-based Nutraceuticals Institute and the FHP is to explore ways of adding value at the agricultural end of the food chain. This could include developing crops with added health benefits, such as vegetables with high concentrations of the phytochemicals that have been scientifically demonstrated as having measurable health benefits, and different ways of processing to maintain health attributes.

Summary of Global Dynamics on the Food System

We characterize the main dynamics of the modern food system below, but it should be noted that our perspective is perhaps a more sober assessment, especially in terms of the neoclassical economic perspective, than is commonly articulated. The modern food system has:

- Witnessed a massive increase in food supply regardless of broader human and environmental health aspects, the economic costs of which are 'externalized';
- Become dominated by certain grains (wheat, maize and rice), and livestock production which promotes meat and dairy product consumption;
- Seen the intensification of agriculture and chemical use with a tendency towards larger production units and fewer crops and farmers;
- Involved costly farm support measures, in the form of subsidies, in the trade-dominating blocs, often at the expense of smaller producers and rural communities and alternative uses of public monies;
- Distorted markets and prompted unequal and unfair trade mainly to the detriment of poorer countries;
- Created a culture of food dependency in developing countries, characterized by 'food aid' and food imports from rich producers and the setting up of domestic production in poor countries for the export markets of the rich food shopper;
- Seen increasing national, regional and global restructuring by large food business and its associated supply industries, built around a select number of commodities (for example Nestlé, Kraft and Unilever are now the world's top three food companies with combined sales of around US$120 billion); and
- Seen environmental concerns (such as falling water tables, reduced biodiversity, soil erosion, chemical contamination and disposal of animal wastes) become major problems.

In this sense the globalization of the food industry is multilayered. Major world commodities such as wheat are now controlled by a handful of often secretive companies. In food manufacturing and processing less than five companies often monopolize particular national markets, for example, in sugar, soft drinks, confectionery, and breakfast-cereal markets. More recently there has been a wave of consolidation and concentration of major food retailers, food-service companies and the food ingredients sector. The food ingredients sector – which supplies starches, flavours, thickeners, sweeteners, stabilizers, colours and so on to food processors, without which the wonders of processed Western foods would be impossible – is indeed the Cinderella sector of the food industry, in the sense that there is scant analysis of its activities. Yet, it is with functional food and healthy eating ingredients that ingredient companies will present food processors with complete 'packages' as solutions for their functional food product needs.

The food industry is faced with consumer challenges, from changing demographic profiles, sedentary lifestyles, a rise in the value of time, increased mobility and higher disposable income on the one hand, to deepening relative poverty on the other. In stark nutritional terms, in the developed world, there are only small increases in per capita intakes in food and these are unlikely to increase in the future.

Writing back in 1981, the Organization for Economic Cooperation and Development (OECD) identified such pressures on resources employed in the food economy and outlined the choices facing the food industry (OECD, 1981):

- Improving efficiency (lowering unit costs);
- Diversification of product ranges and costs;
- Remaining in the food economy, but accepting lower returns on resources than elsewhere;
- Transferring resources to activities providing higher returns; and
- Persuading governments to subsidize the various parts of the food economy.

The OECD choices have proved remarkably prescient. One way that the food industry has responded to the pressures outlined by the OECD has been to embrace health, very reluctantly in the case of healthy eating, but enthusiastically in the case of functional food (see Chapter 3 for explanation). Food, nutrition and health continue to offer the food industry opportunities to wring value and new markets out of slow growth in mature food markets currently estimated at about 1 per cent per annum. The functional foods revolution promises to take health to new heights, or as the scientists openly say, to new frontiers.

Examples of the Impact of Healthy Eating on Food Markets

As we explain in Chapter 3, the earlier healthy eating revolution of reduced fat, fat-free and sugar-free products pointed the way to the immense gains to be had from food and health trends. Let us not forget, for example, that it was as recently as 1983 that the first ever kilogram of the artificial sweetener aspartame was sold in Europe. Today the world market for aspartame is close to 14,000 t and the leading company in this area is NutraSweet (recently sold by its parent company Monsanto).

The consumer goods company Procter and Gamble is reported to have spent more than US$200 million on developing its fat replacer Olestra, which finally won FDA approval in 1996 for use in snack products. The fat replacer was taken up by PepsiCo's snack division Frito-Lay to produce a fat-free potato crisp called WOW! In 1998 WOW! became America's best-selling new brand earning US$347 million in sales, beating other consumer brands from companies such as Gillette, Polaroid and Kodak. 'Fat-reduced' and 'sugar-free' are still big business.

Many in the food industry now regard functional food as the next step, which is why the food industry is in a hurry to find the next 'blockbuster' healthy eating ingredient.

The Challenges Surrounding Functional Food

There are already thousands of products with supposed health benefits available on the world market, and the number of products is accelerating fast. The commercialization of food and health is taking place on an unprecedented scale, using functional food science as a stepping stone. Products range from the nutritionally good to the fraudulent.

Collectively functional foods raise a number of critical questions which are fundamentally different from past healthy-eating products. They include questions about the place of science in product development, the interpretation of risk, disease and illness, marketing communications, regulation and the public health implications of such products. The challenges raised by functional foods can be summarized under four main headings:

1 Food industry challenges. How should products be developed and marketed? In the market case studies detailed throughout

this book, we illustrate how the food industry is currently creating functional food markets.

2 Consumer challenges. While companies are busy market-testing functional food concepts, consumers are also testing whether they really need or want such food products. Key concerns are: can the experts be believed; do functional foods do what they say they do; how much and for how long do you have to eat them to get any health benefit; are they affordable; and do they taste good?
3 Regulatory and policy challenges. Many functional foods and ingredients fall into grey regulatory areas, not least in what can be said in marketing materials and packaging about their health benefits. Also many food and health policy experts are questioning the relevance of functional foods. Many critics of functional foods are concerned that they represent an unwelcome 'medicalization' of food supply. The key debate around functional foods, however, is the issue of health claims and how to regulate them. For example, it is expected that the EU will have some form of health claim legislation in place in less than two years. But how should authorities regulate to allow consumers to make informed choices between good and bad functional foods, and how should they act as enforcement agencies in this area? There is also the challenge of creating a regulatory environment which will stimulate investment and innovation.
4 Scientific and nutritional challenges. There are new scientific developments in the nutritional sciences that have growing and convincing evidence. But what are the implications for nutrition and regulatory policies about food and should recommendations be made to the public? There are other instances where, unfortunately, compelling scientific evidence will not substantiate beliefs in the health benefits of selected foods. The question is often where is the dividing line – in other words just how much science is enough and what should be the response to a lack of scientific proof when this is used in the marketing of food products?

The Market for Functional Food

To date, functional products on the market still contribute limited revenue in terms of total food sales, but their significance to companies far outweighs their current market values. At this stage of market development, market researchers are often vague about what products or product ranges constitute functional food. The best that can be said is that compared to the total food and beverage market, functional foods currently represent a tiny niche. Market estimates are unreliable and should be treated with caution, but as a possible historical benchmark, in 1997, the functional food market was valued at US$1.25 billion in Europe, $14.7 billion in the US and $11 billion in Japan. For comparison, the global food market is estimated to be valued at more than $750 billion. There are other estimates for the size of the functional foods/nutraceuticals market which can vary considerably. One, for example, suggests that by 2010, 25 per cent of European food sales by value could be functional; while for the US it has even been argued that one-half of all food sales are already nutraceutical![10]

In January 2000 market analysts Euromonitor estimated that the world market for functional foods was worth US$27,854 million in 1998. In 1999 sales were predicted to have grown by 13.7 per cent to reach a total value of US$31,661 million, with a substantial increase of 53 per cent in the world market registered over the period 1995–99. In their report, Euromonitor state that the international market for functional food would reach US$51 billion by 2004 if it were not for regulatory interference (Euromonitor, 2000).

Over the past two to three years there has been a remarkable market repositioning by food companies, especially ingredient suppliers, declaring the extent of their commitment to and investment in functional food. Included are companies such as Nestlé, Kellogg, Unilever, ConAgra, Nabisco, Quaker, virtually every major European dairy company (except most UK companies since in functional foods as in other areas of innovation they are the laggards of Europe), DuPont, Monsanto, Novartis (from among life science companies) and many of the transnational food ingredient companies – in fact a virtual 'who's who' of the food industry. Many of these companies have set up functional food or human health divisions to exploit market opportunities.

In short, functional foods are setting public, industry and scientific research agendas; challenging the regulatory systems governing food and drugs; raising concerns among consumer advocacy groups and public health professionals around the globe; and creating marketing and business challenges that have already seen soaring share prices and exceptional profits for some companies. Others have already seen their functional food efforts disappear down a black hole of consumer indifference and hostile media coverage.

For much of the food industry, functional foods present major challenges that it is unused to handling. It is about entering new and uncharted territory for product developers, marketing and for business strategy. Market activity to date suggests that many food companies do not grasp this difference and still see functional food in very much the same light as 'flavour of the month' or the latest product line extension and behave accordingly. Such thinking with respect to functional food is unsustainable – which is why we set out a new business model in Chapter 11.

Using people's fears, worries and concerns about their health and possible future disease risk (in a very different way from consumer concerns about too much fat or sugar in their diet) and even addressing people's already diagnosed medical conditions, as a basis for product development and the marketing of foods on a wide scale is fundamentally different from how traditional foodstuffs are normally marketed and sold. For the food industry it calls for a new approach to product innovation, and the marketing and the communication of food and health issues, not least in terms of ethical marketing. It is a challenge that, in our opinion, few in the food industry are currently meeting in a convincing manner.

In the next chapter we use the real-life story of Benecol, a functional food that at one time came to symbolize the potential of the concept of functional foods. Benecol was seen by many in the food industry as one of the most exciting products of recent years with a potential to change whole food categories. We use Benecol to illustrate the issues and challenges we have highlighted in this chapter and what these mean in practical business terms.

Chapter 2

Benecol – the Rise and Fall of the Colossus of the Functional Food World?

No other product on the world food stage came to symbolize the power and potential of the functional food concept more than Benecol margarine did between 1995 and 2000.

When a certain amount of Benecol is consumed daily, it has been shown in human clinical trials to lower elevated levels of total blood cholesterol by 10 per cent, and more 'harmful' LDL cholesterol by up to 14 per cent. Benecol is seen as the perfect functional food – a proprietary, patent-protected technology, scientifically validated to reduce a near-universally recognized biomarker for the risk of heart disease, all packaged as an everyday food to be consumed as part of a normal diet.

Benecol is also the brand name for the active functional ingredient in the margarine, called plant stanol esters, produced from plant sterols,[1] and developed by the Raisio Group, a Finnish food and chemicals company.

Since its launch in Finland in November 1995 Benecol has come to illustrate the four key challenges facing functional foods:

1 How to develop and, crucially, how to market functional foods;
2 How to gain acceptance by the consumer;
3 Regulatory and policy issues; and
4 Scientific and technology challenges.

Both Raisio and McNeil Consumer Nutritionals, part of the US healthcare giant Johnson and Johnson, and the company that has negotiated the worldwide licensing and marketing rights to Benecol in 1998, have grappled – and still grapple – with all of these challenges in their attempt to create a new functional food market.[2]

Benecol illustrates a further aspect of the functional food revolution – the way investors have seized on the functional food concept as a way of potentially making a lot of money very quickly. From 1995 international investors piled into Raisio's shares, sending their value soaring fifteen-fold before, as disillusion set in, they eventually crashed in late 1999 to a lower price than before the launch in 1995.

What is also remarkable about the Benecol story is that for most of the period from 1995 to 2000, which our case study covers, the main drivers of investor and food industry interest in Benecol were hope and speculative expectation. Up until March 1999, Benecol was only available as a margarine product,in Finland, a country of 5.2 million people. It had been relatively successful, having achieved a 15 per cent value share and 2.7 per cent volume share of the Finnish margarine market – which translates into total retail sales of roughly US$10.5 million (FIM 60 million). Although this is not spectacular by international branded food standards, it sells at roughly five times the price of regular margarine, and is therefore a very high-value niche product.

Benecol's success in Finland was almost certainly helped by two related factors. First, Raisio was already the leader in the Finnish margarine market with a 51 per cent share, so Raisio and its brands already had some equity with the Finnish consumer, from which Benecol could only benefit. Raisio was also experienced in the business of selling margarines. Second, Finland, as we discuss in detail in Chapter 8, as a result of having had one of the highest rates of coronary heart disease in the world, had for more than 20 years benefited from a successful public health education campaign aimed at raising awareness of the links between diet and the risk of heart disease. Raisio, however, do not link the success of Benecol in Finland to the country's public health and nutrition policies in raising awareness about cholesterol and heart health.

Although Benecol was unproven outside Finland, McNeil Consumer Nutritionals reportedly staked US$100 million on its 1999 US launch of Benecol products, pouring its money into all the communication tools of a classic brand launch with the goal of making Benecol a mass-market brand. Such is the power of the belief in a functional food.

There is, however, an important caveat to our case study of Benecol: the story continues as we write. We don't know how the plot will unfold, let alone how it will end – although we will attempt a prediction.

The Development of Benecol

The Benecol story starts in the late 1980s and early 1990s when Ingmar Wester, the research and development manager of Raisio, at that time a virtually unknown company in terms of the global food industry, invented the technology that made Benecol possible. Raisio, tucked away in the south-west of Finland, a few kilometres from the city of Turku, describes itself as a margarine, grains and chemicals group.

While relatively small and not well known in foodstuffs outside Finland, within its home country Raisio is one of the most important food processors. Founded in 1939 by a group of farmers, Raisio is Finland's second largest food company with sales in 1998 of FIM 4,950 million (US$860 million) and is still grounded in its historic origins as a processor of arable produce. It is the largest processor of farm products in Finland, handling grain, rapeseed and potatoes. On average Raisio processes each year around 60 per cent of the total wheat used in Finland, and nearly two-thirds of all potatoes. But with respect to Benecol the salient fact is that Raisio is the established leader in the Finnish margarines market with a 51 per cent market share by retail value.

The idea of using plant sterols as a cholesterol-lowering ingredient was first suggested to Raisio by Professor T Miettenen based at the University of Turku. Finland has a long history of public policy aimed at tackling the country's historically high levels of heart disease, so it was not surprising that a Finnish company saw the potential in cholesterol-lowering concepts. But Ingmar Wester and his research team faced a problem. Plant sterols, the active ingredient found in nature that in certain quantities can block cholesterol absorption in humans, when separated from plants are crystalline and not very soluble, making them unsuitable for use in foodstuffs. Plant sterols are found in relatively small quantities in nature, so another challenge was to find enough of them for economic use in food production. Raisio solved both these problems.

It was already known in the 1950s that plant sterols reduced cholesterol. However, it was in 1988 that university research was launched in Finland to study the effects of rapeseed oil on blood cholesterol, and the Raisio Group was contacted with regard to research on plant sterols. At the same time, Ingmar and his team discovered how to convert plant sterols into plant stanol esters (the result of a process to combine plant sterols with vegetable oils), which enables plant stanol esters to be used as a food ingredient. In 1991 a patent was applied for plant stanol ester and in 1995 the first patents were granted for the manufacture and use of stanol ester. The Raisio Group has patented the stanol ester manufacturing process and the use of stanol ester as a food ingredient in the US, Europe (European patents), Australia, Poland, Russia and Japan. In addition, six new patent families have also been applied for in other parts of the world to expand and extend the patent coverage.

Raisio Goes Public with Benecol

Two events sparked the growth of international interest in Raisio. The first was the publication in the *New England Journal of Medicine* in November 1995 of the results from a 14-month double blind clinical trial using Benecol, which showed that in the group using plant stanol ester, total cholesterol decreased by about 10 per cent and LDL cholesterol 14 per cent compared to a control group (Miettinen et al, 1995). High density lipoprotein (HDL) cholesterol values did not change. The results from the 103 Finns from North Karelia who formed the plant stanol ester group were, it was thought, about to change the lives of the estimated 125 million men and women in Europe and 99 million in North America with elevated blood cholesterol levels who, of course, represented the potential market for Benecol products. Studies have established that the optimal daily dosage of plant stanol (in its fatty acid ester form) is about two grams, which in clinical trials had been shown to reduce cholesterol absorption by 60 per cent. This 2g 'dose' is available from consumption of roughly 25 g of Benecol margarine per day or 40 g of Benecol cream-cheese spreads. The research on plant stanol esters continues and more than 20 different studies have been published. Further clinical trials are taking place in the US, Europe and Japan.

The second event was the actual launch of Benecol margarine in Finland in November 1995. It was reported that the product 'flew off the shelves', despite being four times the price of Finnish butter and six times the price of some regular margarines, and for a time Raisio was barely able to keep pace with demand.

Raisio's Strategy for Benecol

Without resources to roll out Benecol internationally, Raisio opted to license the ingredient and brand name to a global marketing partner, while keeping responsibility for Benecol in its home market as well as, importantly, retaining control of the technology and the production of the plant stanol esters. The licensee would pay licence fees and a royalty on sales as well as buy the active ingredient from them.

One of the goals was to make Benecol the fourth cornerstone of the Raisio Group (in addition to grains, margarines and chemicals) and to this end, in 1998, Benecol operations were concentrated into a new subsidiary, Raisio Benecol Ltd, owned in full by the Raisio Group.

Key to the success of Benecol, the company reasoned, was an adequate supply of raw materials. In 1999, global plant sterol market volume was around 4400 t, of which about half is used by the pharmaceutical industry. Most of the remaining 2000 t (45 per cent of supply) had been reserved by the Raisio Group through various acquistions, contracts and 'call' options. Raisio estimated that some 1100 t of stanol ester would be needed in 1999 for Benecol products, and undertook a multimillion dollar investment in securing stanol ester production, taking stakes in plants on a global basis – from Finland to France, Chile and the US.[3]

According to an estimate drawn up by Raisio and consultants Carta Booz-Allen and Hamilton, the volume of cholesterol-lowering products based on plant sterol would require 6000–8000 t of stanol ester by the year 2005, if the sales targets were met. If the degree of market penetration by Benecol in Finland was also achieved in other Western countries and Japan, demand for stanol ester would surpass 15,000 t by that time.

The International Commercialization of Benecol

In March 1998 Raisio signed a worldwide licensing agreement with McNeil Consumer Nutritionals (then called McNeil Consumer Healthcare) part of the multinational US healthcare group Johnson & Johnson, giving McNeil responsibilty for all Benecol marketing outside Finland. It was hailed by investors and the market as the perfect functional food match, Raisio bringing food expertise and patented technology to the alliance and McNeil a track record in branded health-care products coupled with global reach and impressive marketing muscle.

UK consumers were the first to see Benecol products outside Finland. Launched by McNeil on 29 March 1999, the UK Benecol range consists of regular- and reduced-fat margarines and two cream-cheese spreads. All carried the on-pack claim 'Helps actually lower cholesterol as part of a healthy diet' and are promoted as 'a new tool in the dietary management of cholesterol'. The on-pack claim was later changed to read 'Proven to reduce cholesterol as part of a healthy diet'. Later Benecol yogurts were added to the range. Further launches soon followed in Ireland and in The Netherlands and Belgium, which have a combined population of over 26 million, of whom 60 per cent are estimated to have elevated cholesterol.

At the UK launch of Benecol, Michael Sneed, then managing director of McNeil Consumer Nutritionals Europe said:

> *'There are two reasons why we think Benecol will succeed in the UK. First, the science behind the product is overwhelming and we have spent a lot of time and effort to educate health professionals so that they can be committed behind Benecol. Secondly we are not trying to over-promise what Benecol can do for the consumer. It is not a "be all and end all", but Benecol can help people who are trying to manage their cholesterol.'*

In their marketing, McNeil said, they were at pains to emphasize that Benecol is not a 'magic bullet'.

But in the key US market (where it is estimated over 99 million have elevated cholesterol levels), the Benecol marketing bandwagon was temporarily derailed. McNeil had decided to market Benecol in the US as a dietary supplement, to take advantage of dietary supplement regulations (see Chapter 7) without prior approval from the US FDA, thus allowing stronger health claims to be made than for food products.

The FDA, however, didn't agree with McNeil that Benecol was a 'dietary supplement' – to the FDA, Benecol was a food and should therefore be subject to US food regulation. The practical result of this was a long delay before McNeil were able to introduce Benecol products into the US, holding up the launch by more than six months. Benecol was eventually launched in May 1999 and is now sold as a food product with 'generally recognized as safe' (GRAS) status, but not as a dietary supplement.

In the US, McNeil rolled out Benecol spreads in serving-size packages and, later, Benecol salad dressings. In September 1999 it introduced tub-packed Benecol spreads and three snack products (strawberry, chocolate and peanut bars). The snacks do not require cold storage, paving the way for Benecol's access to thousands of drug stores across the US.

Loss of Competitive Advantage: Unilever Launches Rival Product to Benecol

The global interest in Benecol stirred Unilever, the world's largest margarine producer and third largest food transnational – with sales in 1998 of US$50 billion – into action. Unilever, in the role of market follower, also developed its own cholesterol-lowering margarine, again based on plant sterols. In this case, however, the active ingredient is obtained from vegetable oils such as soy,[4] supplied by Archer

Daniels Midland, one of the world's largest soy processors as well as one of the world's ten largest food companies. The Unilever product is a head-to-head competitor with Benecol.

However, in Europe, where the Unilever product is branded Pro.Activ, Unilever's plans were thwarted by regulatory procedures. Unilever submitted Pro.Activ for review under EU novel foods regulations in December 1998 (Raisio did not have to do this with Benecol since it was launched before the EU novel foods regulation was introduced). Some EU member states raised objections which blocked the subsequent launch – it is alleged that Germany and Sweden had the most serious concerns, suggesting the margarine should be treated as a medicine not a food. It is interesting to note that Finland did not object. Finally, after more than 18 months in the EU novel foods process, the EU Standing Committee for Foodstuffs, comprising representatives from all 15 member states, voted in June 2000 that Pro.Activ is safe following the recommendations, published in April, of the EU Scientific Committee on Food. The first European countries to see the launch of Pro.Activ were The Netherlands and the UK in August 2000. In the UK, Pro.Activ was introduced at a significantly lower price than Benecol – £1.95 ($2.92) compared with £2.49 ($3.73) for a 250g tub of Benecol. UK Pro.Activ packaging says the product: 'Can dramatically reduce cholesterol to help maintain a healthy heart'. In September, Unilever took the cholesterol-lowering spread fight to the heart of Benecol and Raisio with the launch of Pro.Activ in Finland.

In the meantime, in the summer of 1999 Unilever introduced Pro.Activ to the Australian market, where it is sold at four times the price of regular spreads, and in the first seven weeks achieved a 7 per cent value share and 2 per cent volume share of the Australian spreads market. The product was launched in New Zealand in October 1999. Unilever should have enjoyed a significant first-mover advantage in these markets, but within a few months Goodman Fielder – Unilever's main rival in the region – had rolled out its own plant-sterol-based cholesterol-lowering margarine, under the Logicol brand, demonstrating just how difficult it can be to maintain any kind of truly unique advantage in the functional food marketplace.

Unilever also experienced the same delays as McNeil/Raisio in the US, seeking to introduce their brand 'Take Control' under dietary supplement regulation. But in the end the Unilever brand came to market just ahead of Benecol in May 1999, but at a considerably lower price.

The Bubble Bursts

Although Benecol was launched into the international markets in 1999 on a huge wave of expectation, by the end of the year it was already apparent that those expectations were not going to be met. In December 1999, the US trade magazine *Advertising Age* broke the story that McNeil was pulling out the TV portion of its US$100 million campaign to market Benecol in the US after the hoped-for level of sales had failed to materialize. It was this report which pricked the Benecol bubble, sending Raisio's stock diving 30 per cent in one day – Raisio's 'black Thursday' (*New Nutrition Business*, January 2000) – and leaving the company's shares trading at less than a quarter of their autumn 1998 peak.

Figures also came to light from Information Resources Inc (IRI), the Chicago-based product tracking and research company, which revealed that in the period from launch to 10 October 1999, Benecol sales were just US$13.7 million – $8.2 million for Unilever's competing Take Control brand – giving predicted annual sales of about US$25 million. Sales of Benecol salad dressings were so low that they did not even register on IRI's radar screen. By the summer of 2000, Benecol sales had hit the $30 million mark, but importantly, sales of Unilever's Take Control had nudged ahead with sales of $31 million.

In early 2000 Raisio was to reveal just how little penetration Benecol had achieved in its launch markets. The following are average market shares for Benecol margarine, measured by value as a percentage of the total consumer spreads markets: Finland 12.3; UK 3.2; US 2.8; Belgium 1.5 and The Netherlands just one per cent. However, by July 2000, Benecol products seemed to have gained a significant foothold in the UK marketplace. Trade sources estimated Benecol was selling £400,000 worth of product a week, making the UK a £20 million market on an annualized basis. Importantly, McNeil was seeing a high level of repeat purchases.

Benecol's Poorly Conceived International Strategy

Benecol was, in theory at least, a highly differentiated product because of a claim which was unique, clinically-proven and targeted at an important indicator of heart disease which could be used on-pack and in product marketing. The claim would, it was believed, act as a significant signal of value, appealing to consumers' growing interest in the links between their diet and their health and so justifying a significant price premium.

But the challenge for any company with a differentiated product is actually finding the price premium which will maximize returns while still getting consumers to buy the product. In the Finnish market Raisio had set the price of Benecol at five times that of regular margarines and Benecol had become a highly successful, but niche, product. McNeil adopted the same niche-pricing strategy, but then set off to market Benecol as a mass-market product, putting a reported US$100 million behind US promotion alone.

It is one of the basic rules of marketing that price premiums are only acceptable to buyers when a product's differentiating features are valuable to the buyer. The lesson of Benecol is that a cholesterol-lowering claim – Benecol's key, indeed only, differentiating feature – is not valuable enough to create, other than for a very small group of buyers, a value five times that of regular products. The premium which consumers will accept, it is now known, must be somewhere below that level. But finding it is the challenge for food marketers. It is not surprising that, at the time of writing, Raisio had announced a review of Benecol pricing.

One of the main assumptions underlying Benecol was that its cholesterol-lowering message was in some way unique. But this belief does not stand up to close examination. In the UK, for example, Benecol sits in the chiller cabinet among a 'wall' of other yellow fat products. One of its neighbours is the Flora brand (sold in mainland Europe as Becel), a spread which carries the statement, 'as part of a healthy diet helps lower cholesterol', on its packaging – a statement which Unilever, the owners of 'Flora', have scientifically established. To the average consumer, comparing this to Benecol's claim that it 'helps actually lower cholesterol as part of a healthy diet', a significant difference between the two claims is probably not immediately apparent. The consumer might also notice on the 'Flora' pack a heart-shaped logo and the words 'approved by the Family Heart Association', while Benecol carried no such third-party approval as a signal of value. What, the consumer could be forgiven for wondering, justifies Benecol being four to five times the price of 'Flora'? Moreover, 'Flora', a brand which has been around since the 1960s, has since the 1970s been cleverly marketed on a 'heart-healthy' platform under the 'Flora Project', a holistic nutrition marketing effort, while being priced at the same level as other spreads.

It is of course the role of marketing communications to help differentiate a product and build value in the mind of the consumer. But, as we will discuss in Chapters 10 and 11, messages about functional foods are often complex and difficult to convey. Some of the most successful food and health communications rely primarily on long-term and often quite sophisticated consumer education rather than, for example, 30-second TV advertising slots. In the case of Benecol the advertising and public relations campaigns clearly failed to create an impression of product value in consumers' minds which was sufficient to justify the high price premium. By the end of 1999 McNeil had discontinued its US TV advertising in favour of focusing on communications to physicians, which had been a key component of the Benecol communication strategy since its launch.

Marketing to physicians may, ironically, reinforce Benecol's position as a niche brand. Many medical doctors have poor knowledge of nutrition. They receive limited training in this area and many members of the medical profession are sceptical or even hostile to the concept of prevention through nutritional means (Heimburger et al, 1998; James, 1997b; Temple 1999). Health professionals may, in fact, have only limited impact on nutritional advice to the population at large.

We can illustrate our point of view by way of a true, but of course anecdotal, story. A 69-year-old woman, who knew she had moderately elevated levels of cholesterol, on seeing the advertisements on UK television in 1999 went out and bought Benecol. Shortly afterwards, during a scheduled visit to check on her progress, she told her physician about being 'on the Benecol'. The physician at this time knew nothing about the product, but wrote it down in her patient notes, and told her to 'keep on taking it'. Incidentally, he didn't view her cholesterol condition as needing medication. Over the coming months, our everyday senior citizen found it impossible to comply with 'taking' Benecol three times a day. She said she only managed to eat it once a day and also, living on retirement income, she found it expensive to buy. Before long she gave up and, as she told us, 'went back to Flora'! On a further visit to her physician he asked if she was still taking Benecol and she informed him she wasn't. The physician said nothing, and did not encourage her to start taking Benecol again. He also offered, as on the previous visit, no nutritional or dietary advice or suggestions as to where she could get such 'preventative' nutrition help. The woman did notice that in the physician's waiting room there was now a pile of Benecol promotional leaflets, but, she told us, she didn't bother to take one. Obviously this is a one-off case, and perhaps it is not at all representative (but imagine if it was!). The point is that nowhere in this anecdote is either the woman or doctor emotionally engaged with Benecol as a brand. The experience was clinical and 'medicalized'. Emotion is one of the most powerful drivers of brand loyalty and many successful brands are successful in part because they become emotionally connected with their consumers.

Another factor which may have undermined Benecol was the need to consume two or three servings a day in order to get the cholesterol-lowering effect. As one major retailer explained to us, this was just not convenient enough for many buyers, suggesting that the ideal functional food should offer – at maximum – a daily dose to be effective.

Packaging too, is one of the most important ways in which a product's value is signalled to the buyer, but the packaging of Benecol margarine is not particularly eye-catching. It does not stand out among all the other similar-looking tubs of margarine and it does not clearly signal that it is really different. By contrast the packaging of Goodman Fielder's Logicol cholesterol-lowering spread, launched in Australia, uses both colour and type of packaging – very different from other spreads – to project a high-quality, 'expensive' image.

At this stage of the functional food revolution many companies believe that the ability to use a claim will be vital to the success of such foods and there is a restless search to develop products with unique scientifically-validated claims that will earn high price premiums. The case of Benecol throws doubt on this assumption.

An Untried and Untested Regulatory Gamble

It was the problem of health-claims regulation which Raisio themselves blamed for the poor performance of Benecol, referring to what they call the shortcomings of legislation on functional foods. In particular they point to the regulatory problems they experienced in the US that delayed the initial launch of Benecol products.

At first glance such an explanation seems very plausible. But if we look closely at what McNeil actually did it can equally be seen as a strategic error on their part which, with hindsight, they now want to blame on regulations and policy makers. Rabbe Klemets, the head of Raisio's Benecol division, was quoted in the *Financial Times* as saying: 'The message has been blurred. We cannot go out and say "this product does such and such for you"' (25 February 2000).

But the regulatory excuse has to be seen for what it is. Under current US legislation, food and drugs, for understandable reasons, are clearly and separately defined and regulated. It is a regulatory framework every company faces. Many companies see the lack of regulatory clarity to make so-called health claims as holding back the market for health-promoting foods. But this has not stopped many companies – from Kellogg to Campbell's – from developing functional foods. In fact many are succeeding without using health claims. In the

UK, for example, McNeil use the claim 'actually lowers cholesterol' on Benecol packaging and in advertising, but despite this clear message sales were still disappointing.

The truth is, McNeil tried to launch Benecol in the US as a dietary supplement under the Dietary Supplement Health and Education Act (DSHEA) 1994 – a move that was largely untested. They were therefore taking a massive regulatory risk which backfired on them. DSHEA allows so-called structure–function health claims for dietary supplements (see Chapter 7), but whether DSHEA could be used for foods is a very grey area, subject to much debate, although a few small companies in minor markets had introduced food-like products as dietary supplements (for example, Basics Plus, a kefir – a drinking yogurt-style product).

Benecol, and the way it was going to be marketed, would have completely changed the concept of foods as dietary supplements in the US and, potentially, opened the floodgates for mainstream foods to reposition themselves as dietary supplements. It is common and open knowledge that the US FDA are concerned about this last point. Also, it should be remembered that the passing of DSHEA was a major defeat for the FDA – they were not going to watch a range of everyday foods becoming dietary supplements. McNeil, therefore, initially attempted to take Benecol to market using an untried and untested regulatory method for the introduction of an important category of food products.

It should also be remembered that the regulatory wrangle only delayed the US launch of Benecol by around six months (something that also affected their main competitor Unilever) – the disappointing level of sales after launch was a product of a poor pricing and marketing strategy, not the regulatory situation.

Managing the Investment Community and the Stock Analysts

The frenzy which surrounded Raisio's share price in the period from 1995 to 2000 is one of the defining characteristics of the Benecol story. Indeed Raisio was the subject of media and investor hysteria[5] of a type which, at the end of the 20th century, was characteristic of Internet stocks but hitherto unknown among food companies. Table 2.1 illustrates some of the key market rumours affecting Raisio's share price, while Figure 2.1 shows the 15-fold increase and subsequent collapse in Raisio's shares. Figure 2.2 details how foreign ownership of Raisio increased from under 10 per cent before the launch of

Table 2.1 ***The Raisio/Benecol Share Price Chronology***

1995	Raisio launches Benecol margarine in Finland
1996	According to market rumour the First Boston Investment Bank evaluates Benecol patents as worth FIM 11 billion. Raisio doesn't comment; shares start steep climb
1997	Raisio announces licensing agreement with Johnson & Johnson to market Benecol in the US. (One market analyst stock market note claims that Johnson & Johnson paid Raisio an initial payment of US$10 million, followed by a further US$50 million, with an agreement that continues to pay Raisio a percentage royalty based on gross sales). Sharp jump in share price
1998	According to market rumour Johnson & Johnson would be buying Raisio. Surge in Raisio share price. Becomes public knowledge that Unilever is developing its own cholesterol-lowering margarine as competitor to Benecol. Halts surge in share price. Raisio announces results from a study undertaken at Kuopio University, Finland, that confirms Benecol's cholesterol-lowering properties. Share price continues climb to 1998 peak. According to market rumours Unilever ready to launch rival product earlier than expected.
1999	FDA blocks marketing of Benecol in US as dietary supplement which prevents making strong packaging health claims. Delay of US launch. Unilever's European launch put on hold after delays caused by EU novel foods process. In April Benecol given GRAS status and clearance for sale in US. Raisio warns that Benecol profits will be much less than expected. Press reports that Benecol sales in trouble in US; Raisio issues stock market press release on 2 December 1999 and share price plunges almost 30 per cent in one day.
2000	In January Raisio announce they are reviewing marketing strategy with McNeil. In May Unilever get all-clear to launch their competing product Pro.Activ in Europe.

Benecol to almost two-thirds as investors speculated on the potential of Benecol.

Investors were taken with the idea that Raisio had the advantage of a unique, patent-protected technology which would translate into blockbuster cholesterol-lowering products. Stock market analysts made predictions for Benecol which were based more on hope than analysis, driving up expectations of future profits and making Raisio's share price highly volatile. In the spring of 1996, for example, investors piled into Raisio after the First Boston Investment Bank

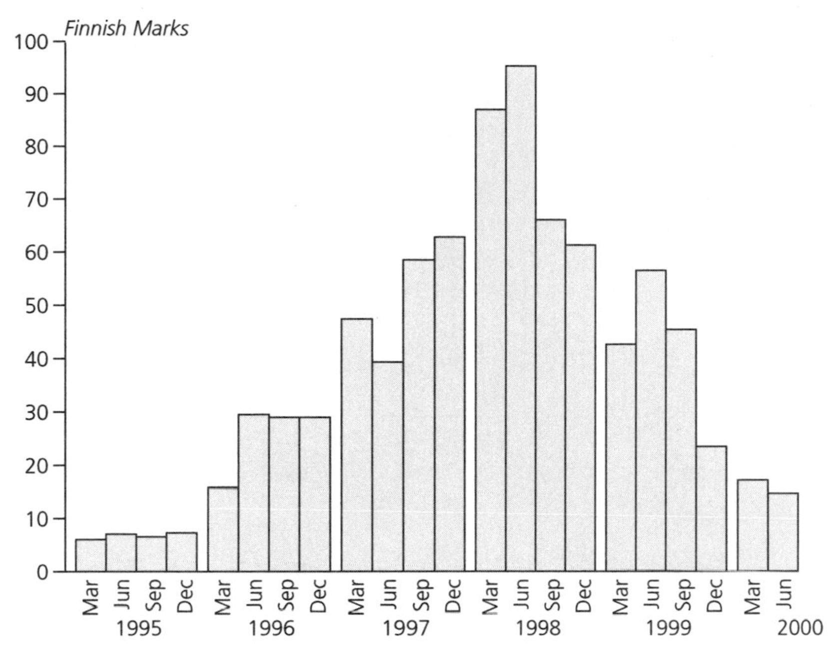

Note: Mean rate Raisio Group plc, Free Share split adjusted, June 26 1998
Source: Raisio, 1999

Figure 2.1 ***Raisio's Share Price at Selected Dates (1995–2000)***

made what Raisio saw as a wild estimate of Benecol's patent value, putting it at US$ one billion. A London analyst was quoted in 1996 as saying 'Benecol could be generating profits of FIM 1.2 billion by 2000, based on a 3 to 5 per cent share of the developed world's margarine market' (*New Nutrition Business*, 2000). The shares again surged in early 1998 on the unfounded market rumour that Johnson & Johnson were going to purchase Raisio.

In fact the idea that Benecol's 'uniqueness' was sustainable because it was patent-protected was soon proven false with the introduction of similar products from Unilever and then from Australia's Goodman Fielder. Analysts were also apparently ignorant of the fact that Archer Daniels Midland, one of the world's largest food processors, had set up a business specifically to sell plant sterols produced under its patent – a clear signal that many more cholesterol-lowering products would be coming to market.

As a result of this international attention, ownership of Raisio stock by non-Finnish investors soared from less than 10 per cent in 1995 to nearly two-thirds in 1998 and remained at around 50 per cent

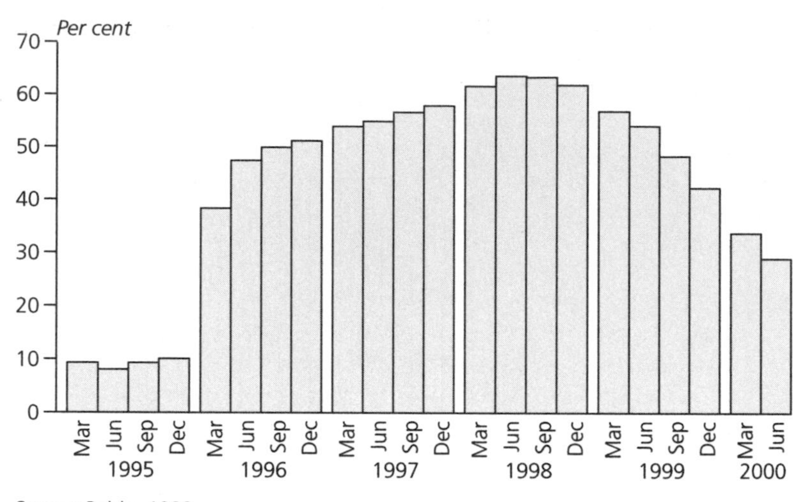

Source: Raisio, 1999

Figure 2.2 *Raisio's Foreign Ownership at Selected Dates (1995–2000)*

in September 1999. In an interview reported in Finnish newspaper *Helsingen Sanomat* Tor Bergman, at the time Raisio's chief executive officer, explained that the balancing act of reconciling the views of foreign owners with those of domestic owners, the latter predominantly Finnish farmers, was often difficult because of very different priorities. Raisio's farmer shareholders expected the company to look after its traditional strengths – such as the processing of wheat and oats – while foreign investors, Bergman said, were only interested in Benecol. Raisio products such as french fries and flour, he added, didn't raise much interest among US pension funds and other large investors. The foreign owners, Bergman said, would want to take Benecol alone to the stock market and abandon uninteresting or old-fashioned businesses.

Foreign speculation in Raisio, however, has made some senior managers very rich – the so-called Benecol millionaries. Under a share option scheme introduced in September 1993, ten managers in Raisio have done very well from foreign investment. For example, Matti Salminen, Raisio's chief executive officer before Tor Bergman took over in the summer of 1999, made FIM 14 million (US$2 million), while it is estimated Tor Bergman himself made FIM11–12 million (US$1.8 million). Under a new share option scheme, introduced in 1998, more than 100 Raisio employees stand to gain if the share price performs at levels above FIM100. The slow start to Benecol means that for them the hopes of gain seem a very distant prospect.

To complicate matters, in 1998, Raisio's traditional business began to have problems of its own. For example Raisio had 20 per cent of the Russian margarine market – worth FIM 800 million – but this business literally collapsed overnight with the devaluation of the Russian rouble in August 1998. As a by-product of the collapse of the Russian market, Raisio became very dependent on Benecol for its profits.

In 1998, Raisio's Benecol division had a turnover of about 6 per cent of total Group sales rising to about seven per cent in 1999. But, in the same periods, the Benecol subsidiary contributed nearly 60 per cent of total Raisio Group profits in 1998 and 46 per cent in 1999.

McNeil's failure is proving doubly expensive for Raisio who have invested heavily in plant sterol ester production – the Benecol ingredient. In 1999, only 20–25 per cent of Raisio's stanol ester raw material capacity was in use and its operations were unprofitable.

Another important lesson from Benecol is the importance of corporate openness and honesty. The fall in Raisio's share price on 'black Thursday' was compounded by a 'day of silence' from Raisio's top management. According to press reports in Finland, investors were furious by what some Helsinki stockbrokers called the 'invisible' leadership of Raisio. Tor Bergman said that the company would be reviewing their communications policy.

It was partly in response to this perceived need for greater openness that Raisio admitted some of their mistakes and set out their future plans in a Situation Report (see Raisio website) on 21 February 2000 and in a presentation to analysts in London, described by the *Financial Times* on 25 February as 'sombre, almost penitential' (p18).

So Where now for Benecol?

Raisio's response to the disappointing start to Benecol's internationalization was to announce at the beginning of 2000 that it would take over responsibility for marketing Benecol in the Baltic region (Sweden, Norway, Denmark, Poland, Russia and Iceland), where the company already had distribution, in addition to its Finnish home market. In an apparent sideswipe at McNeil, Raisio were reported as saying that Benecol products in these countries would have been launched two years earlier if they had been in control then. Raisio says it is also working actively on a market launch in Germany and Japan.

In addition, Raisio are to seek out third-party users for Benecol. Based on the concept of 'Intel inside' used in the computer industry – or the NutraSweet ingredients branding used on, for example, soft drinks – they hope the Benecol 'sub-brand' could be used to expand the range of Benecol-type products from dairy, confectionery and drinks. The goal is to use Raisio's plant stanol ester raw material capacity more fully by such sales to third parties. The company has set itself a goal of sales for the Benecol business of FIM 1200 million (around $180 million) by 2005 with operating profits of 20–30 per cent of turnover and is now talking in terms of making Benecol a success over a five- to seven-year period – a timescale which, we argue in Chapters 10 and 11, is more appropriate to the challenges posed by the functional foods revolution. The lesson of 1999, however, is that in order to be successful Raisio may need to exercise some control or influence over the marketing of Benecol-based products and/or invest in its own marketing of Benecol ingredients.

Benecol products, however, will not be the only new cholesterol-lowering foods in the market. Apart from Unilever, Novartis has already launched a cholesterol-lowering product (part of its Aviva range) which is to be marketed in both Europe and the US, and other companies have cholesterol-lowering ingredients and products in the pipeline. The concept of cholesterol lowering is being taken up by many competitors. The most recent group to position their brands under the cholesterol-lowering banner are soy food producers, who in October 1999 secured an FDA-approved health claim for soy in relation to heart disease, based on soy protein's potential to lower cholesterol (see Chapter 7).

The Benecol story illustrates the key factors driving functional foods:

- The new ways nutrition science and functional food science are being commercialized and used to develop food products;
- The global character of the marketing and technology;
- The focus on making an everyday food product a part of disease prevention;
- The regulatory and policy difficulties these products present; and
- The commercial significance of functional foods to the food industry.

In addition, the Benecol story illustrates the misunderstanding of functional foods markets by investors and analysts. Food markets by tradition are conservative and tend to change over the long term (making food stocks 'defensive'). Functional foods are not 'blockbuster' drugs and to succeed in them companies will have to plan for the longer term.

The Benecol story is compelling and the lessons are many, not least in the use and development of nutrition science and patented technology expertise to create differentiated and new market opportunities. The science and clinical trials behind Beneol are outstanding and a model for the industry. However, Benecol challenges the widespread conception that consumers are willing to pay premiums of four and five times the price of regular products for foods targeted at specific diseases.

It may also be, as we discuss in Chapter 11, that McNeil have relied too heavily on what we describe as a 'medical model' of functional foods to develop and promote Benecol. What is more this approach is backward looking, stuck in the healthy-eating revolution paradigm we outline in the next chapter.

While Raisio have a unique technology, in one sense they were hardly 'first'. Food and health has been a dominant global food market trend for more than 15 years. There is a wealth of material on the marketing of food and health from which Raisio and McNeil could have learnt a great deal. It therefore seems less than credible that Raisio and McNeil were unaware of the difficulties they were up against and unable to plan for them – just look to Japan with thousands of functional food product introductions, or the European dairy industry with hundreds of functional food products, where, ironically, Finland's largest dairy company has been a pioneer. Closer to home, Raisio and McNeil could have studied the clever marketing of Unilever and the way they developed their heart-health platform in margarine going back to the 1970s.

In the next two chapters we take up the themes of what characterizes the healthy-eating and functional food revolutions, especially as these have become, or are attempting to become, institutionalized within public policy. It is in this broader, public policy environment that functional foods will either flourish or wither. As the example of Unilever illustrates, with its entanglement with public policy in Europe delaying the planned launch of Pro.Activ, and McNeil's strategic error in trying to launch Benecol as a dietary supplement in the US, public policy can have a devastating impact on business development in functional foods. So while policy issues are, by their nature, often time-consuming, protracted and appear divorced from the real world, for food and health and particularly functional foods, time spent understanding and devising strategy that accommodates such concerns can mean the difference between business success or failure.

Part 2

Risk, Regulation and Policy Dilemmas

Chapter 3

A Clash of Two Nutrition Revolutions and Cultures?

One of the many facets of the functional foods revolution is that it cuts a swathe across policy related to food and health. Successful businesses will have to understand the policy environment in which they operate and relate creatively to this in the context of their own product and business development. To understand the policy environment pertinent to functional foods, we distinguish between the healthy-eating revolution and culture, and the functional foods revolution.

By the healthy-eating revolution we mean the widespread government promotion throughout the developed world, starting in the 1970s, of dietary goals and guidelines outlining dietary change for healthy adults in relation to a number of diseases and illnesses, but in particular heart disease. The central premise of the healthy-eating revolution, with minor country by country variations, is to eat a variety of foods; maintain ideal weight; eat foods with adequate starch and fibre; avoid too much sugar; avoid too much sodium; avoid too much fat, saturated fat, and cholesterol; and to drink alcohol in moderation. The healthy-eating revolution is built on a remarkable consensus among scientific and nutrition experts.

Making a sharp distinction between the healthy-eating revolution and the emerging functional foods revolution is a novel approach to analysing nutrition and public policy in relation to food and health. Our analysis therefore should be seen as exploratory, as a tool to highlight areas for future investigation and a means of characterizing the key elements and dynamics of this aspect of food policy. The goal is to set the stage for an informed debate among the numerous stakeholders and interests necessitated by the functional foods revolution. Our thinking in laying out such an agenda is our belief that a major policy and cultural conflict is emerging between the functional foods revolution and its ensuing commercialization, and the still recent, but earlier healthy-eating revolution, which played out from the late 1970s onwards.

Unless this conflict is reconciled we can envisage one scenario in which functional foods could be relegated to minor, niche markets and effectively banned from general food supply. As far as many proponents of the healthy-eating revolution are concerned, there is already a scientific consensus on what a healthful diet should be. Functional foods, by emphasizing a narrow range of dietary components and their potential impact on disease and often relying on specific ingredients, are from this perspective seen as 'technical fix' or 'magic bullet' solutions that detract from wider public health goals related to nutrition. But the functional foods revolution presents challenges to both industry and public health policy. It moves beyond healthy eating into new and uncharted territory when seen from a marketing and public policy perspective.

Not surprisingly, in its short life the concept of functional food has attracted both controversy and condemnation (Milner, 1999). In part this is the result of the policy implications of functional food, although this is rarely articulated or analysed. In general the dynamics and culture of the functional food revolution is in stark contrast to the healthy-eating revolution that evolved during the 1980s. This is expanded on in detail in the rest of this chapter, the contrast between the two being summarized in Table 3.1.

Definitions of Policy Jargon

As with much policy or other specialist debate, there is a great deal of jargon involved and complex nuances to address. So to make it clear what is being discussed, some straightforward definitions are called for, in particular of the terms food policy, nutrition policy, public health and public policy.

By public policy is simply meant 'policy made by government or government agencies' (Germov and Williams, 1999). Of special significance for functional foods are regulatory aspects of public policy, most notably because they often fall in a grey regulatory area between food and drug (Heasman and Fimreite, 1998). One of the key areas of activity in the public policy sphere, is the development of health claims for functional food. This aspect of functional food and public policy is covered in detail in Chapter 5. In addition, the analysis presented in both Chapters 6 and 7 on Japan and the US is perhaps surprising in that it demonstrates the extent and role of public policy in shaping functional food markets and marketing.

Nutrition policy and food policy are often used interchangeably, but we, like many others see these as quite distinct. Food policy, for example, is the study of policy in the context of the whole food system, that is, the set of activities and relationships that interact to

Table 3.1 *Contrasting Characteristics of the Healthy-Eating and Functional Foods Revolutions*

Healthy Eating	*Functional Food*
Substantive scientific consensus	Mixed scientific validation and extent of knowledge
Public health policy push	Public health policy resistance
Extensive food industry resistance	Extensive food industry push
Widespread 'official' promotion	Largely commercially promoted
Focus on changing total diet and 'balance'	Focus on 'hit' products/ingredients
No 'good' or 'bad' foods	'Good' foods now available
Public policy dietary interventions	Market-led dietary interventions
Standardized population/individual based dietary goals	Confused dietary goal messages
Long-term evaluation/monitoring	None or limited evaluation in place
Authoritative sources of dietary advice	Experts seem to disagree or contradict each other
Public health models prominent	'Medical' model dominates
Media plays crucial role	Media still plays crucial role
Set stage for functional foods	Builds on 'healthy eating' revolution
Consumer skepticism/poorly informed	Consumers better informed/want information
'Healthy eating' sells products	Functional foods still to prove themselves

determine what, how much, by what method and for whom food is produced. This set of activities involves a complex set of relationships, operations and transformations of inputs which ultimately result in food being presented to consumers. For the sake of clarity, food system is used to mean broadly the same as food economy or food chain (OECD, 1981). Nutrition policy, as an element within food policy, is defined as 'sectoral policies concerned with food supply that have an explicit mandate to take the health and well-being of the whole population into consideration' (Helsing, 1997).

By this definition, nutrition policy is also clearly an element of public health. Put simply, public health is an assessment of health needs and services aimed at the population level. Public health is best distinguished from clinical medicine by its emphasis on preventing disease rather than curing it, and its focus on populations and communities rather than the individual patient (Bloom, 1999). Public health concerns take us into such aspects as considering the burden and economic cost of disease and illness on society and the societal conditions which promote their development; therefore health is not simply a personal choice, but also reflects processes at work in the wider society that require a public response (Stein, 1995).

However, much public health and nutrition policy focuses primarily on individuals and their habits or lifestyle when defining health and nutrition issues and the means to address them. This is in contrast, for example, to the environment that frames the array of options available to individuals and from which they, for good or ill health, choose patterns that become their lifestyles (Milio, 1990). Such a public health debate – that is the environmental versus the individualist approach – has become a central feature of the arguments over the implementation and implications of the healthy-eating revolution (for a global perspective see Lang and Heasman, 2001; for a UK example see James, 1988; for US public health policy examples see Sims 1998; see also Milio 1993 and Hesling 1989).

The Healthy-Eating Revolution

Healthy eating today is so commonplace and widely known that it seems incredible that it is the result of major policy battles and controversies that raged from the late 1970s and for most of the 1980s (Cannon, 1997, Sims, 1998). Even today some of the basic tenets of the healthy-eating revolution are still being challenged (Willett, 1997). But it is the healthy-eating revolution that has set the stage for functional foods, particularly in how it has educated consumers and policy makers in the benefits of diet and health and how nutrition policy can benefit public health. In addition, parts of the food industry have learnt how to turn the healthy-eating revolution into new and profitable marketing and product development opportunities that have successfully created massive new markets, while meeting consumer expectations and improving consumers' food choices. It is worth lingering briefly over how such a radical change in perception came about and the lessons to be learnt for functional foods.

As the food system changed rapidly in the 20th century, increasing concerns were being raised about food, diet and public health. Initially the main concern was with deficiencies in the diet – that is, an insufficiency of the essential nutrients that result in illness if not supplied by the diet. The science of nutrition itself originated with the discovery of essential nutrients such as the various vitamins and essential amino acids. The concept that disease would be caused by an excessive intake of nutrients (over-consumption) increasingly caught the attention of researchers only after the Second World War, when these problems became more widely known through the work of researchers such as Ancel Keys and others (Keys, 1953 and 1980). These researchers were trying to understand the origins of the heart disease epidemic that had struck the US even before the War (Milio and Helsing, 1998). The suggestion, highly controversial at the time, was that diet could play an important role in heart disease.

But as Milio and Helsing point out, while studies like this were received with some attention, in contrast to the continued interest generated by deficiency diseases, there was: 'a certain resentment against the idea that too much of a nutrient might be bad for health' (Milio and Helsing, 1998). At this time high-fat diets came under suspicion as the possible causes of certain diseases. But – importantly – Milio and Hesling say it did not take long for either nutritionists or food producers to realize what this might mean in terms of food, and the potential loss of economic growth in certain food-producing sectors.

The diet-degenerative diseases link was dealt what seemed like a final blow in 1962, when a World Health Organization (WHO) expert committee stated:

> *'It must be repeated that at the present time there are no effective means by which the occurrence of ischaemic heart disease can be prevented. Such therapeutic measures as are available must therefore be applied to the disease itself in the hope of delaying its progression or preventing late complications... Nevertheless, much further research is needed before public health authorities can recommend major alterations in the diet, or are justified in advising that more or less of any particular kind of fat would be beneficial.'* (Milio and Hesling, 1998)

With hindsight, and in the light of all the subsequent attention given to the kind of fat in the diet, this reads as a truly remarkable statement today. In summary, starting in the 1950s, but not gaining widespread acceptance until the 1970s, the post-war diet in the developed world came to be seen as a public health issue. One of the key policy differences was the change in emphasis from micronutrient deficiency to one where food itself became associated with chronic disease. The need for corrective nutrition policies in relation to human health began to see unprecedented government intervention in attempts to shape the diets and food consumption patterns of the Western world – the healthy-eating nutrition revolution was upon us.

The Core of the Healthy-Eating Revolution

By the 1990s, the central pillars of the healthy-eating revolution had become something of a nutritional mantra, so that by the end of the 1980s commentators like Scrimshaw were able to write, 'After years of controversy, a remarkable degree of consensus has developed regarding the kind of nutritional goals most likely to promote good health' (Scrimshaw, 1990).

However, the core recommendations of the healthy-eating revolution most likely to promote good health, have dramatic, if implicit, economic implications. If followed as a whole, they define a distinct dietary pattern. When translated into food choices, this pattern derives most daily food energy from grains, vegetables and fruits, with less energy from meat and dairy foods, and even less from fats (Nestle, 1999).

Truswell details very succinctly the evolution of the healthy-eating revolution through dietary goals and guidelines (Truswell, 1987; see also McNutt, 1981 and Callaway, 1997 for a historical perspective on dietary guidelines for Americans; and Sims, 1998 and Kritchevsky, 1998 on recommendations about dietary fat). Truswell is careful to distinguish between recommended daily allowances (or intakes) and dietary goals or guidelines. Recommended daily allowances are the levels of intake of essential nutrients (for example vitamins) considered by expert committees, on the basis of available scientific knowledge, to be adequate to meet the known nutritional needs of practically all healthy persons. As we illustrated in our earlier example (described in Chapter 1) about new recommended vitamin C intakes for Americans, functional food science is also offering new challenges for some RDAs.

Dietary goals or guidelines are more recent than RDAs and aim, not to provide enough of the essential nutrients, but to reduce the chances of developing chronic degenerative disease. Dietary goals and guidelines form the scientific underpinning of what we mean by the healthy-eating revolution. Importantly, they are aimed largely at a healthy, adult population and are therefore preventative in character. The commercialization of the healthy-eating revolution is expressed in the thousands of fat-reduced, low-fat, fat-free, sugar-free, sugar-reduced and high-fibre product launches in the food and health marketplace since the 1980s, a trend which continues to this day. For example in the US, on average, more than 1000 new low-fat and fat-free products have been introduced annually since 1990 (Sims, 1998). Most of these foods are designed to taste identical to their full-fat counterparts.

Low-fat Foods Reach an All-time High in the 1990s

Research by the US trade magazine *New Product News* shows that low-fat foods made up almost 16 per cent of all food products introduced in the US market in 1996, reaching an all-time high with 2076 new products in that year (Table 3.2). This was in a year when total new food product launches were down by 21 per cent. By and large, the products making low- or reduced-fat claims were more of the same, such as entrees, ice cream, salty snacks, bakery products and yogurt.

Table 3.2 *Numbers of US Products Bearing Nutritional Claims, (1989–96)*

Product	*1989*	*1990*	*1991*	*1992*	*1993*	*1994*	*1995*	*1996*
Reduced/low-calorie	962	1165	1214	1130	609	575	1161	776
Reduced/low-fat	626	1024	1198	1257	847	1439	1914	2076
All natural	274	754	661	996	449	575	407	645
Reduced/low salt	378	517	572	630	242	274	205	171
No additives /preservatives	186	371	526	631	543	251	167	143
Low/no cholesterol	390	694	711	677	287	372	163	223
Added/high fibre	73	84	146	137	51	26	40	12
Reduced/low sugar	188	331	458	692	473	301	422	373
Added/high calcium	27	20	15	41	14	23	21	35

Note: Health claims categories are not additive as new products may carry more than one claim
Source: New Products News, 1997

The Internationalization of the Healthy-Eating Revolution

The first set of dietary goals was published in Sweden in 1968 and formed the basis of Scandinavian policy interventions such as the Swedish Diet and Exercise campaign and the Norwegian Nutrition Policy. Table 3.3 lists a number of expert committees recommending a change in diet for the general population throughout the 1970s and early 1980s. By the 1970s there were two lines of reasoning for dietary advice to the public. Truswell (1987) says: 'The Scandinavian nutrition professors aimed primarily to replace *empty calories* (Truswell's emphasis) with a variety of nutritious foods while the followers of Ancel Keys aimed primarily to bring down the plasma (total) cholesterol concentration.'

The First Report on *Dietary Goals for the United States*

If there was one report that sparked off the healthy-eating revolution it was the first edition of *Dietary Goals for the United States*, published in 1977. This report was put together by a group of politically interested activists, but the involvement of Dr Mark Hegsted of Harvard University ensured there were no major nutritional follies. The report was notable because it was among the first to bring together several diet–disease hypotheses for which the scientific evidence had started to mount – such as the importance of dietary fibre, the possible relation of high fat intake to breast and bowel cancer, and the relationship between hypertension and salt intake. In addition it set quantitative target levels for reducing fat, saturated fat and cholesterol in the American diet (Sims, 1998).

Truswell describes the report as a 'revolutionary document'. The document created a storm of protest and objections and polarized professional opinion. However, many of the principal US researchers on diet and atherosclerosis approved. By the end of 1977 a second and revised edition had been published and this included a dissenting foreword with the following statement:

> *'The value of dietary change remains controversial, and science cannot at this time insure that an altered diet will provide improved protection from certain killer diseases such as heart disease and cancer.'*

Table 3.3 ***Expert Committees Recommending a Change in Diet for the General Population***

Country/Organization	*Date*	*Committee/Report Title*
US	1970	Inter-Society Commission
New Zealand	1971	Royal Society
New Zealand	1971	Heart Foundation
US	1972	American Health Foundation
US	1973	White House Conference
US	1973	American Heart Foundation
US	1973	American Heart Association
UK	1974	COMA-Department of Health and Social Security
Australia	1974	National Heart Foundation
Australia	1975	Academy of Science
Norway	1975	Report to the Storting
New Zealand	1976	Heart Foundation
UK	1976	Royal College of Physicians and British Cardiac Society
Canada	1976	Department of National Health and Welfare
US	1977	Dietary Goals (first edition)
FAO/WHO	1977	Dietary Fats and Oils
US	1977	Dietary Goals (second edition)
Canada	1977	Quebec Nutrition Policy
US	1978	American Heart Association
Australia	1979	Department of Health
Australia	1979	Association of Dietitians
US	1979	Surgeon General
US	1979	American Society of Clinical Nutrition
US	1980	US Department of Agriculture/ Department of Health Education and Welfare
France	1981	Apports Nutritionnels Conseilles
Australia	1982	National Heart Foundation
US	1982	American Heart Foundation
US	1982	National Research Council Committee on Diet, Nutrition and Cancer
WHO	1982	Prevention of Coronary Heart Disease
New Zealand	1982	Nutrition Goals
UK	1983	NACNE
UK	1984	COMA-Department of Health and Social Security

Source: Based on Wheelock, 1986

Among the criticisms were the following: more research was needed; relationships were not provided; it was politically motivated; it promised the public too much; it was unreliable to base recommendations on food disappearance date; it was based more on intuition than scientific reasoning; diets should be individually prescribed by the medical profession; it was puritanical; it smacked of the big brother approach; and it was a nutritional debacle (Truswell,1987).

However, while never an official document, it remained in circulation and drew public, as well as professional attention, to the need for guidance on diet and health. For the first time, the consumer was being called upon to make not only quantitative decisions – how many servings of food to eat – but qualitative choices which meant eating less of some foods and more of others. As in other countries, the terms dietary goals and dietary guidelines are both used. Generally, dietary goals are expressed in nutrients (for example, types of fats), while guidelines are expressed in foods, such as 'eat plenty of grain products, vegetables, and fruits', although in practice dietary guidelines also include goals. A further differentiation is that goals offer quantitative dietary targets, while guidelines offer qualitative advice.

According to Sims, the report served to crystallize scientific opinion between those who believed in a targeted approach, aimed at people who had been identified as being at high risk, and those who believed in a public health approach – that is, having the general population adopt health-promoting behaviour, including changes in diet, that would do no harm, but might be able to prevent the early onset of particular chronic diseases. The relevance of this debate is a central concern to the way that the functional food revolution will unfold and in many instances will have a direct input into marketing strategy. Remember that a key characteristic of functional foods is that, as currently commercialized, they target two populations – on the one hand healthy adults and on the other those with a recognized medical condition, disease or illness, a distinction which is consistently confused in marketing practices.

In effect by early 1980 the public health approach to nutritional advice was institutionalized with the publication of *Nutrition and Your Health: Dietary Guidelines for Americans* by the US Department of Agriculture (USDA) and (as it was then) of Health, Education and Welfare. (The *Dietary Guidelines for Americans* are reviewed and published at five-yearly intervals, with the latest edition published in 2000; see Box 3.1).

Support for the Healthy-Eating Revolution

Between 1961 and 1991 more than 100 authoritative scientific reports on food, nutrition and public health were published throughout the world (Cannon, 1992). In a later article Cannon summarizes what he calls 'a well-established consensus among the scientific community that the case against the Western diet is proved beyond reasonable doubt' (Cannon, 1995; see also Wheelock, 1992). While drawing attention to the fact that some reports are more cautious than others, and that in these the case is proved on 'the balance of probability', Cannon expresses what he regards as the general agreement among published reports on nutrition and public health in the following way:

- During the last half-century, Western diets have become unbalanced. They now contain too much fat in general, too much hard, saturated fat in particular, too much sugar and salt, and not enough fibre;
- Translated from nutrition to food, a healthy diet is rich in vegetables and fruit; bread, cereals (preferably whole grain) and other starchy foods; and may include fish and moderate amounts of lean meat, and low-fat dairy products; and
- The best diet to reduce the risk of heart attacks is the best diet to protect against obesity, diabetes, common cancers and other Western diseases, and is also the best diet to promote general good health (Cannon, 1995).

Dietary guidelines vary considerably in the amount of background documentation provided with one of the most thoroughly documented being the US National Research Council's 750-page report *Diet and Health* published in 1989, with 33 experts on the main committee and another 76 specialists providing input (Truswell, 1998). Truswell summarizes the report's headline recommendations as:

- Reduce total fat intake to 30 per cent or less of total energy. Reduce intake of saturated fatty acids to less than 10 per cent of energy and that of cholesterol to less than 300 mg/d;
- Eat five or more servings of a combination of vegetables and fruit, especially green and yellow vegetables and citrus fruit, daily. Also increase intake of starches and other complex carbohydrates by eating six or more servings of a combination of breads, cereals, and legumes daily; and
- Balance food intake and physical activity to maintain appropriate body weight.

Six other recommendations deal with protein, alcohol, salt, calcium, supplements and fluoride.

Dietary guidelines are assumed to apply to healthy adults (Callaway, 1997), and are based on an epidemiological or population-based models. They assume an equal risk of disease, given the identification of certain characteristics or risk factors, such as age, gender, race, body weight and behaviour such as smoking (the concept of risk is discussed in more detail in Chapter 4).

As already outlined, government has become heavily involved in the acceptance and promotion of healthy-eating advice. But this is in itself controversial, with a strongly held view among some that government should stay out of the business of giving advice to the public about what to eat. One aspect of this point of view is that the responsibility for behaviour change lies with the individual, not with government (see Sims, 1998).

Healthy-eating guidelines have not changed very much over time. For example, in 1997, in the UK, the Health Education Authority published a guide entitled *The Balance of Good Health* for nutrition educators in which healthy eating had been simplified to eight general dietary guidelines for the public:

1 Enjoy your food;
2 Eat a variety of different foods;
3 Eat the right amount to be a healthy weight;
4 Eat plenty of foods rich in starch and fibre;
5 Eat plenty of fruits and vegetables;
6 Don't eat too many foods that contain a lot of fat;
7 Don't have sugary foods and drinks too often; and
8 If you drink alcohol, drink sensibly.

It is interesting to note that not a single interpretation of the new science or the concept of functional food plays any role in government advice about healthy eating as we start the 21st century.

The Politics of Healthy Eating

The consistency of healthy-eating advice over time also shows that experts do not disagree on the essential messages of good health through food and diet, contrary to some impressions (Nestle, 1994).

But the healthy-eating revolution has met with massive resistance from major sections of the food industry and from agriculture, particularly livestock producers, dairy farmers, sugar producers and related agribusiness interests (Knutson et al, 1990). The revolution has profound implications for food purchases. For example, to consume the recommended healthy-eating diet, people must reduce their intake of the principal food sources of saturated fat and cholesterol – meat, dairy products and eggs – and avoid processed foods that are major sources of fat, sugar and salt (Nestle, 1994).

When the US government began to publish dietary recommendations for chronic disease prevention, certain sections of the food industry responded by lobbying the USDA and Congress to discredit, weaken or eliminate any advice to reduce purchases of their products (Milio, 1991).

Even in Scandinavian countries with advanced and integrated nutrition policies, industry resistance has been fierce. For example, Norway regards an integrated food and nutrition policy as an important part of the overall effort within society to prevent disease and promote health. This in turn must have a close connection with agricultural, fishery, price, consumer and trade policies as well as educational and research policies. However, the Norwegians themselves point out that the changes taking place have not been as marked as they wished and the rate of change has taken longer than expected. Norum (1997), for example, says progress in implementing the nutrition policy was slow in the first years after its approval by the Norwegian parliament in 1976:

> *'Two main reasons were that the dairy and meat industry were against the policy, and that the National Nutrition Council had little power and political influence. The dairy industry tried to counteract the policy by producing foreign experts who claimed milk, butter and other dairy products had no influence on risk factors for coronary heart disease, and that therefore the Norwegian nutrition policy was built on false premises.'*

The Politics of the Healthy-Eating Revolution in the UK

The UK is given as an example of the slow implementation of healthy-eating nutrition policy (Cannon, 1988; James, 1997B). As in the US, it was the publication of an unofficial nutrition report that brought the healthy-eating revolution to public attention. Wheelock (1986) provides an insightful account of the background to the publication

Box 3.1 The 2000 edition of *Dietary Guidelines for Americans*

In May 2000, President Bill Clinton announced the release of the fifth edition of the *Dietary Guidelines for Americans*. The new guidelines continue to emphasize balance, moderation, and variety in food choices, with a special emphasis on grain products, vegetables and fruits, and have been improved to be more consumer-friendly. The dietary guidelines serve as the basis for US nutrition policy and provide advice to consumers about food choices that promote health and decrease the risk of chronic disease. It is a legal requirement for the dietary guidelines to be updated every five years to incorporate advances in medical and scientific research.

In these latest guidelines, there is advice on keeping food safe to eat, particularly the need to keep and prepare food safely in the home. In addition, the guidelines focus on physical activity as important for healthy living, more than just for weight management. For example, the guidelines point out physical activity can help build and maintain healthy bones, muscles, and joints; build endurance and muscular strength; and promote psychological well-being and self-esteem.

The language of diet

As Table 3.4 shows, the language, if not the central themes, used in the US dietary guidelines has changed significantly over the past 20 years. The latest edition, perhaps most dramatically than most.

For example, the new dietary guidelines represent a major change in presentation. They introduce three basic messages for the first time: Aim for fitness; Build a healthy base; and Choose sensibly – for good health (the ABC's for good health). The intent of these messages is to help the user to organize the guidelines in a memorable, meaningful way.

Other key changes in language used are an emphasis on 'daily' routines for the first time, for example, being physically active and choosing grains, fruits and vegetables 'daily'. Another subtle, but significant change, is the inclusion of choosing 'beverages' to moderate sugar consumption – an acknowledgement of the vast quantities of sugar consumed through soft drinks by some Americans. Also new is the inclusion of choosing 'whole grains' on a daily basis. Another dietary practice change in the new guidelines is pointing out to use less salt when 'preparing' foods, another first for the guidelines.

Public consultation on the guidelines

The development of each addition of the guidelines involves extensive public consultation and lobbying of the Dietary Guidelines Advisory Committee that is set up to produce the new edition. Four public meetings were held by the committee in Washington, DC in September 1998, March, June, and September 1999. Written comments from the public were received throughout the process. During the March 1999 public meeting 40 organizations provided oral testimony. Meeting transcripts and the final report of the committee are posted on the Internet at www.ars.usda.gov/dgac.

A total of 165 submissions with recommendations from public sources were made to the Advisory Committee drawing up the new Dietary Guidelines. Public comments were received from individuals, interest groups, industry, academia, state and federal government agencies and elected officials.

Table 3.4 *Language of Five Editions of* Dietary Guidelines for Americans *(1980–2000)*

1980	*1985*	*1990*	*1995*	*2000*
Maintain ideal weight	Maintain desirable weight	Maintain healthy weight	Balance the food you eat with physical activity – maintain or improve your weight	Aim for a healthy weight – be physically active each day
Avoid too much fat, saturated fat, and cholesterol	Avoid too much fat, saturated fat, and cholesterol	Choose a diet low in fat, saturated fat, and cholesterol	Choose a diet low in fat, saturated fat, and cholesterol	Choose a diet that is low in saturated fat and cholesterol and moderate in total fat
Eat foods with adequate starch and fiber	Eat foods with adequate starch and fiber	Choose a diet with plenty of vegetables, fruits and grain products	Choose a diet with plenty of grain products, vegetables, and fruits	Choose a variety of grains daily, especially whole grains. Choose a variety of fruits and vegetables daily

1980	*1985*	*1990*	*1995*	*2000*
Avoid too much sugar	Avoid too much sugar	Use sugars only in moderation	Choose a diet moderate in sugars	Choose beverages and foods to moderate your intake of sugars
Avoid too much sodium	Avoid too much sodium	Use salt and sodium only in moderation	Choose a diet moderate in salt and sodium	Choose and prepare foods with less salt
If you drink alcohol, do so in moderation	If you drink alcoholic beverages, do so in moderation	If you drink alcoholic beverages, do so in moderation	If you drink alcoholic beverages, do so in moderation	If you drink alcoholic beverages, do so in moderation

Source: Adapted from Sims (1998) and *New Nutrition Business* own analysis

Summary: 2000 Dietary Guidelines for Americans

Aim for fitness

- Aim for a healthy weight
- Be physically active each day

Build a healthy base...

- Let the Pyramid guide your choices
- Choose a variety of grains daily, especially whole grains
- Choose a variety of fruits and vegetables
- Keep food safe to eat

Choose sensibly

- Choose a diet that is low in saturated fat and cholesterol and moderate in total fat
- Choose beverages and foods to moderate your intake of sugars
- Choose and prepare foods with less salt
- If you drink alcoholic beverages, do so in moderation

of this report, which is generally known in the UK as the NACNE report. The National Advisory Committee on Nutrition Education (NACNE) was established in the late 1970s under the aegis of the

British Nutrition Foundation and the Health Education Council. Very early on NACNE realized that it would not be able to make progress on its remit to bring nutrition education up to date until it had guidelines on what actually constituted a healthy diet. As a result the committee asked Dr (now Professor) Philip James to assemble a group with the object of preparing guidelines. Wheelock writes that it was an open secret that the committee under James had difficulty in reaching agreement on dietary recommendations. He says:

> *'There are strong indications that pressure was exerted by certain industries, who considered that sales of their products would be adversely affected if the recommendations led to changes in food consumption patterns.'*

Despite the initial conclusions from NACNE being first presented to the Department of Health and Social Security (DHSS) in 1981, the DHSS had continually refused to accept them (Wheelock, 1986). Then in July 1983, an exclusive article in the UK national newspaper *The Sunday Times*, broke the NACNE story and the views of a group of experts on what was wrong with the British diet was out (Cannon, 1987).

The DHSS responded to these allegations and issued a statement to set out the background to the establishment of NACNE. The DHSS made it clear that NACNE was not an official government committee and had no remit to advise government on nutritional standards. However, it was agreed to publish the final version of the James Report (as it also became known) in October 1983.

Although *The Sunday Times* article (written by Geoffrey Cannon) did not state specifically that an attempt had been made to suppress the NACNE report, the article had, according to Wheelock, a significant impact and 'many people honestly believed that the DHSS was actually delaying the report because of pressure from the food industry'.

The impact of publication of the NACNE report was very similar to that of the publication of the first edition of the *Dietary Goals for the United States* in 1977. The NACNE report also detailed for the first time in the UK quantitative dietary goals for the population, but the essential dietary changes proposed were at the time highly contentious – but are the now familiar:

- reduction in total fat consumption, especially saturated fat;
- increase in polyunsaturated fatty acids consumption;
- reduction in sugar consumption;
- increase in fibre consumption; and
- reduction in salt consumption;

An essential part of the NACNE recommendations was institutionalized by the DHSS less than a year later, on 12 July 1984, by the publication of the Committee on Medical Aspects of Food Policy report (COMA) on diet and cardiovascular disease. Also like NACNE, the COMA report made quantified dietary goals, but in the case of COMA the focus was for reductions in fat consumption in relation to cardiovascular disease and not healthy eating *per se*. The report and its recommendations were accepted by the government of the day, thus becoming official nutrition policy for the UK. The COMA report also included a recommendation for nutrition labelling on foodstuffs as a way to educate the public about nutrition.

However, as James points out (James, 1997) there was widespread academic criticism in the UK of the dietary basis of coronary heart disease with the establishment seemingly opposed to the idea it could be caused by diet. He cites by way of example that the British Heart Foundation systematically rejected all attempts to develop a prevention policy and rarely funded work on dietary aspects of coronary heart disease. Similarly, the UK's two cancer research bodies, the Cancer Research Campaign and the Imperial Cancer Research Fund, were adamant that diet had little to do with the development of cancer (James et al, 1997b). James argues that Britain lags a long way behind other industrialized countries in seeing diet as intrinsic to public health – an attitude for which he holds doctors largely responsible. He has been reported as saying: 'The medical establishment finds it hard to believe that food and nutrition have much to do with public health.' Doctors, James says, simply have not taken on board modern nutritional concepts or the fact that moderating diet can sometimes have a greater impact than drugs (White, 1998).

Despite population-based dietary goals and guidelines being endorsed as public policy, the main intervention by governments in nutrition policy in Western countries has been through labelling and education for the individual. For example, Wiseman (1990) writing about nutrition policy in 1980s-Britain, says the general philosophy of the government of the time was very much that responsibility lay with the individual. He explains that this means in practice that individuals should be able to make informed choices for their diets. In order for any choice to be informed, there must be enough information available and the individual must be educated to interpret that information usefully. Thus Wiseman concludes 'the output of nutrition policy is essentially the provision of information via labelling and education'.

But the provision of information has been characterized in some quarters as an unequal contest between the limited public funds devoted to healthy-eating promotion and the massive advertising and promotional budgets of certain parts of the food industry. For example, in the UK, the Health Education Authority received £700,000 funding specifically for the nation's health education for 1996–97, while in 1995, according to the UK Advertising Association, £551 million was spent on food and drink advertising (Keane, 1997). Confectionery is the most heavily advertised food category, followed by coffee, fast foods and soft drinks. The majority of advertisements on UK children's TV are for food and drinks; breakfast cereals are the most heavily advertised, followed by confectionery, fast foods, soft drinks, ice cream and lollies (Food Commission, 1994).

Healthy-Eating Arguments Descend into Abuse

In the UK, at times, debate over nutrition policy and food and health has degenerated from reasoned arguments to trading terms of abuse. For example, Passmore (1985) describes the enthusiasts who urge UK consumers to change their dietary habits as detailed in the NACNE report as 'food propagandists' and writes:

> *'I have no sympathy with these new puritans who would diminish our enjoyment of eating...the puritans who wish to alter the quality of foods that we have enjoyed for years and who advocate widespread dietary changes require much more hard scientific evidence to support their claims...'*

Reports, such as *A New Diet of Reason* (Conning, 1995),[1] continued to debate the merits of dietary guidelines in the UK saying, for example, that simple dietary guidelines – the currency of public health pronouncements – are unlikely to achieve more than reduce excessive consumption of food or address inadequate intake of some dietary components. Conning neatly captures the individualistic point of view (in contrast to the public health perspective), when he writes:

> *'Diets do not cause disease. The diseases we have been assessing (arterial disease, cancer,* [diseases of] *the nervous system) arise through the normal processes of ageing allied to inherited tendencies and are not the result of being poisoned by individual foods or nutrients, even in excessive amounts. The causes of the diseases are not known but some of the processes involved are susceptible to modification by several means, one of which may be dietary change.'*

However, Lang reminds us that the 'contested space' of food policy (including nutrition policy) in the UK was not solely the remit of non-governmental organizations (NGOs) or individual activists. Those contributing to the debate came from many quarters, including agricultural, nutritional, technological, feminist and consumer interests (Lang, 1997). It also included many within the food industry. However, the broad base was not acknowledged and the debate often descended to personal attacks from some opponents. For example, in a much-quoted phrase, John Gummer, the Minister of Agriculture in the late 1980s and early 1990s, referred to the need to counter 'food fascism'. His junior parliamentary secretary at the Ministry of Agriculture, Fisheries and Food in 1992 referred to 'food terrorists' and the director of the right-wing Social Affairs Unit (the publishers of *A New Diet of Reason*) talked of 'food Leninism' (all quoted in Lang, 1997).

In countries such as the UK, Sweden and Ireland, where physicians and nutritionists disputed the evidence on diet and health, James (1997a) points out there was little change in cardiovascular disease rates. He goes as far as to say that: 'This provides a clear example of how conservative thinking can itself be held responsible for thousands of premature deaths.'[2]

The Functional Foods Revolution in the Context of the Healthy-Eating Revolution

For many professional nutritionists trained and working within the culture and consensus of the healthy-eating revolution, developments in functional foods are often seen as alarming and even deeply disturbing. In the UK, for example, there has been a hardening of views against functional foods among many health professionals and nutritionists. At the same time, consumer advocacy organizations, often champions of healthy eating, appear cynical of functional food developments. A typical response is captured in this extract from a letter to the *British Medical Journal* (Jacobson and Silverglade, 1999):

> *'Fortifying conventional foods with physiologically active substances might sometimes be appropriate, but the unbridled marketing in the United States of about $12 billion worth of dietary supplements annually shows the potential for defrauding and sickening consumers. The spread of such mischief to the far larger*

food industry could prove disastrous. The composition of, and advertising claims for, functional foods should be governed by judicious government regulations, not by corporate marketing strategies.'

The concept of functional food as complementary to healthy eating and the selling of this idea and the science supporting it to the policy community is a major challenge for the food industry and the scientific community. At the same time, while many in the policy community are at times selective in the examples they use to condemn functional foods in general (unfortunately some companies are producing the products to fuel this continued scepticism), the policy community is also trying to understand the real scientific developments in food and health and the implications of these for consumer and possibly public health. In short, we have little idea whether functional foods are becoming a cause for concern, or for celebration, when it comes to long-term public health.

The scale, scope and enormity of the functional foods revolution is only just starting to emerge, as will become apparent through the market and business developments detailed throughout the rest of this book. As with any new and evolving science and market, the concept of functional food begs many questions, whether it is over efficacy, over health claims, how products should be regulated or how these foods fit into the total diet. In addition, consumer advocacy organizations are expressing a number of concerns as, for example, in the title of a recently published report which sums up their perspective: *Functional Foods – Public Health Boon or 21st Century Quackery?*

This report title begs the question whether functional food will contribute to public health goals or simply be seen as addressing a narrow range of unsubstantiated individual effects. But a clearly stated objective of functional food science is to contribute to human, and by implication, public health. Where appropriate there is a public responsibility to provide objective information to consumers on the role of diet in health, including new developments. There is also a frustration among a number of food companies who have pioneered the science and research and development in functional food, that they are unable to effectively communicate their science-based health findings to consumers. Consumers are therefore unable to differentiate between responsible marketing and opportunist companies with little or no scientific integrity.

The Choices for Public Policy and Functional Food

There are three possible ways forward for public policy in respect to functional food. The first, to put it simply, is to resist what is happening. There are strong merits to this position. For example, there are already clearly established dietary guidelines for health and these are a long way from being fully implemented – why, some will question, do we need functional food? In addition, tough regulations could ensure that health claims are only allowed when a predetermined 'gold standard' of science is achieved.

Second is a half-way house, a wait-and-see approach, giving a little here (on health claims for example) and urging caution in other areas. The danger of this approach is that it will be largely reactive and that it could exaggerate confusion in the public mind. This seems to be the current policy response.

The third approach – a radical response – would be for public policy to wholly embrace the functional foods revolution. This would mean developing a more proactive food and health policy to optimize nutrition in the sense implied by functional food science. For example, the food industry is in effect looking to promote 'good' foods, often at the expense of 'bad' foods. Public policy should look to make strong health claims for the former.

Public policy could also take the functional foods revolution further. The essential cornerstone for embracing the functional food revolution should be what we call the 'A+' approach to food and health, with the 'A's standing for 'availability', 'affordability', 'accessibility' and 'acceptability'.

The 'A+' Approach to Functional Food and Health Policy

Functional foods should be:

- available – on shelves where people normally go shopping;
- affordable – within the means of most people;
- accessible – for the benefit of the general population; and
- acceptable – it doesn't matter what the science may show, if functional foods are not accepted in the minds of consumers they will fail.

In his 1997 Caroline Walker Lecture, Professor Philip James said the challenge is to engage the major food companies as allies rather than obstacles to public health (James, 1997). The functional foods revolution could be interpreted as offering such an opportunity. The confrontational politics of the healthy-eating revolution, as highlighted in this chapter, is not a realistic or constructive way forward for the functional foods revolution. Functional food is clearly a policy issue that embraces public, nutrition and food policy as well as public health policy, and successful food companies will not be able to ignore their responsibilities in these areas. But functional food also gets to the heart of medical research, in particular the communication of the risk of disease in relation to dietary habits. The functional foods revolution takes the notion of risk in relation to diet to new areas of policy and food marketing. In the next chapter we consider this critically important area in more detail.

Chapter 4

Eating and Drinking in the Risk Society – Implications of Risk for Functional Food

Eating is an extremely risky activity. The US National Food Safety Initiative attributes 9000 deaths and between 6.5 million and 33 million episodes of illness annually to food-borne microbial illness (Sanders, 1999). Further, diet is a contributing factor in four of the ten leading causes of death in countries such as the US – the four being, heart disease, some cancers, cerebrovascular disease, and diabetes mellitus (Busch and Williams, 1999).

But the functional foods revolution thrives on risk. Its central premise is that the risk of disease can be curtailed by dietary components, from the selected consumption of individual phytochemicals derived from plants to eating live bacteria that can colonize the human gut and impart health benefits.

Let us take as an example a box of the toasted whole grain oat breakfast cereal Cheerios, made by General Mills in the US. Emblazoned on the front of the cereal box, across what appears to many reasonable observers to be a heart-shaped bowl, are the words 'Cheerios may reduce the risk of heart disease'. The message is clear. The reduction of risk in relation to health is the big idea behind functional foods and nutraceuticals. In fact a healthy lifestyle has become almost synonymous with a lifestyle characterized by risk evasion (Forde, 1998).

Why the Concept of Risk is Important to Functional Food

One of our reasons for highlighting the concept of risk is that at times it is being interpreted incorrectly from medical and other studies to apply to general populations in the development of functional food products. Nowhere is this confusion more apparent than with the many food marketers' lack of concern with the distinction between relative and absolute risk, which we discuss later in this chapter. We have personally heard food company representatives discuss the results on risk reduction from studies done with sick people as if these translate easily to wider, healthy populations and cultures. For functional food, therefore, the notion of risk is central to:

- the construction of health claims;
- the development and use of biomarkers or endpoints;
- communication and marketing strategies about food and disease prevention that do not mislead the public; and
- the place of functional food in public health policy.

The idea of risk and disease and what these mean to ordinary people, is a central issue for functional food, no matter what perspective is taken. The whole premise of the functional foods revolution is that by consuming certain food components, the risk of disease is somehow reduced or that diseases will even not occur. But just what does risk prevention mean and how should this risk be translated to the everyday foods we eat, or, more relevantly, how should functional foods be constructed or developed and marketed to optimize risk reduction?

The concept of risk and the modern food supply is a vast and complex issue. It raises many difficult methodological, technical and policy problems that need addressing further if functional foods are too succeed and become a useful part of the diet.

Before we turn to risk, food, and disease prevention, we briefly examine the wider context of risk, not least to introduce how risk is socially constructed. This concept encompasses not only how ordinary people understand health risks in relation to their diet, but also how policy makers, politicians and scientists interpret risk in relation to food and diet. In many respects, the social aspect of risk is largely ignored in scientific studies of food components and disease, not through any conspiracy, but because it is not necessarily the role of science to consider it. Looking at the social construction of risk highlights two often contradictory concerns relevant to functional food. The first is how to communicate new and positive scientific findings about nutritional risk and second, how to address the potential abuse of the idea of risk in the presentation and marketing of functional food to consumers.

What is Risk?[1]

At first glance, the problem of risk in society seems relatively clear cut. Risk simply brings together the notion of a hazard or danger with calculations of probability (Lupton, 1999). Probability and thinking about probability are key concepts in understanding risk.

Lupton (1999) neatly summarizes the technical-scientific debates about risk as:

- focusing on how well risk has been identified or calculated;
- the level of seriousness of a risk in terms of its possible effects;
- how accurate is the science that has been used to measure and calculate risk? and
- how inclusive are the causal or predictive models that have been constructed to understand why risks occur and why people respond to them in certain ways?

It is then a short jump to see the solution of risk problems in society as the resolution of conflict between scientific, industrial and government organizations and the public in relation to the health and environmental risks associated with science, technology and industry.

In the EU there has been a rush to employ the language and action of risk assessment and risk management as the cornerstone of European food policy. This follows a series of damaging food scares in Europe, including Britain's BSE crisis, the Belgian dioxin-in-food and evidence of sewage sludge being made into animal feed, to name just three prominent examples. To oversee this risk management approach to food policy it is proposed to set up a new European Food Agency by 2002.

However, a weakness of this perspective is an underlying assumption that ordinary people's view of risk is inappropriate or incorrect and as responding unscientifically to risk and can therefore be corrected through appropriate risk management and risk communication. Assumptions like these have been challenged by research. For example, Powell and Leiss (1997) concluded that the 'public carries a much broader notion of risk' than do most researchers and physicians. The public's notion of risk, they say, includes 'accountability, economics, values and trust'.[2]

The confusion over risk and its complexity as socially constructed is nowadays no better illustrated than over the protracted and increasingly confrontational arguments about the introduction of biotechnology in the food chain. It is here that some of the more worrying implications of the social construction of risk can be seen at work in relation to the traditional technical-scientific concept of risk. This aspect of risk and GM food has been summarized by the UK-based ESRC Global Environmental Change Programme in a paper (1999) on risk, science and public trust. The programme's work challenges a number of widely held assumptions and includes insights ranging from how ordinary people actually demonstrate a well-informed understanding of issues related to new technologies to questioning the validity of the widely held notion of sound science on which policy makers often rely.

The social construction of risk can play an important part in the priorities adopted in policy making. An example from the UK can illustrate the point of how sound science on risk can be risk-managed in the policy area. For example, while some scientists argued that the risk of contracting BSE from eating beef-on-the-bone was minimal, politicians saw fit to ban its consumption, justifying the decision on public health grounds. On the other hand, the scientific community has reached some degree of consensus that increased consumption of fruits and vegetables can reduce the risk of some cancers by 30 per cent in the general population. Therefore, lack of fruit and vegetables seems a much greater current risk to human life than BSE from eating beef-on-the-bone, but the political resolve to meddle with food supply to ensure vastly improved access to fruit and vegetables is lacking.

Risk and the Food Supply

Sanders (1999) lists a range of risks associated with food with different degrees of severity in the developed and developing world:

- nutritional deficiency;
- nutritional imbalance (for example, obesity, excess intakes of salt, saturated and *trans* fats)
- naturally occurring toxicants in food (for example, alkaloids, legume toxins, cyanogenic glycosides);
- microbiological contamination (bacteria, viruses, parasites, mould and algal toxins)
- contaminants in food (heavy metals, organic chemicals).

To these could be added other food hazards (Lang and Heasman, 2001), which include, but are not limited to:

- widespread use of agrochemicals (such as pesticides);
- food processing ingredients (such as certain food additives);
- food and agricultural production methods (for example, the use of hormones as growth promoters in beef cattle; widespread use of antibiotics in livestock production);
- food industry handling and hygiene practices;
- the introduction of novel processes and technologies (food irradiation; biotechnology in agricultural production).

It is no exaggeration to say that the conflicting meanings, understandings and applications of notions of risk have become one of the central concerns of the global food economy. These range from risk assessment and risk management strategies for food safety hazards, including risk disputes centred on the trade in foodstuffs, to the introduction of new technologies and novel processes and, our focus in this chapter, risk in relation to disease reduction and the diet we eat.

Perceptions of Risk

Perhaps of paramount importance, from both a business and a health promotion point of view, is how people view ideas about risk and their health. Following on from this question is how people translate their concerns about food and health into their dietary choices. This area is poorly researched in relation to the concept of functional food (for a detailed discussion of models of consumer-perceived risk from a marketing perspective, see Mitchell, 1999). As in the clash between the healthy-eating and functional food nutrition revolutions, a central debate about risk and human health is the explanation of individual risk in relation to population risk.

Assessing Risk as a Basis for Health Claims

The role of risk in experts' assessment of scientific literature that may be presented as the basis for a health claim for a food product is a highly technical, but nevertheless critical, issue. Health claims are a recurring theme throughout this book, but without appropriate risk assessment in terms of disease reduction there would be no basis for health claims. The important consideration here is how experts construct their idea of risk and also the appropriate form of risk communication. For example, as we point out in our review of the US in Chapter 7 the wording of many health claims, decided by expert consensus based on studies that assess the reduction of disease risk, is often far from consumer-friendly (Hasler, 1996).

But there appears to be little work on the construction and perception of risk by experts working on aspects of food supply. This seems like an important area for further investigation, as Lupton (1999) says, in describing one perspective on risk:

> *'A risk...is not a static , objective phenomenon, but is constantly constructed and negotiated as part of the*

> *network of social interaction and the formation of meaning. Expert judgements of risk, rather than being the "objective" and "neutral" and therefore "unbiased" assessments which they tend to be portrayed in the technico-scientific literature, are regarded as being equally constructed through implicit social and cultural processes as are lay people's judgements.'*

Research in a number of areas (but not specifically on food) shows that scientists clearly recognize that all risk assessments are inherently uncertain and subject to change when new data are obtained (Stayner, 1999). The problem, Stayner points out, is that all stages of the risk assessment process (hazard identification, dose-response analysis, exposure assessment, and risk characterization) are fraught with uncertainty, which frequently leads to acrimonious debates among scientists and others about whether there is a risk and about how best to quantify it. He also reminds us that 'risk assessment has become a battleground for powerful interests that are potentially affected by regulation'.

Expert disputes over risk are not always readily acknowledged in discussions about risk communication and how these might influence the nature of risk communication. For example, Sanders (1999), in his thorough overview of food production and food safety, writes:

> *'A major barrier to risk communication is a general lack of understanding by the public of relative risk as opposed to absolute risk. Furthermore public perception of risk is distorted by media reporting.'*

Statements such as these could be construed as one-sided in the sense that they do not acknowledge or address the disputes over risk communication between scientists, the role played by vested interests in furthering these, and how these expert disagreements influence consumer and media perceptions. It continues to perpetuate the one-sided point of view that in the social construction of risk, it is ordinary people who are at fault and that it the expert who understands the sound science of risk. As we shall show, in the medical profession there has been a dramatic about-turn in guidelines on relative risk as opposed to absolute risk in the prevention of heart disease; in other words, the understanding of relative risk as opposed to absolute risk is not just an issue for the public.

The functional foods revolution holds the potential for the food industry to make major blunders by unintentionally – and in some cases deliberately – incorrectly communicating risk and disease prevention in support of health claims for products. It is convenient, though not so clear-cut as implied by Sanders, to lay blame at the door of media reporting. We live in an age dominated by information and media of all kinds; the media is a fact of modern life, yet its plurality and competition means it cannot afford just one point of view. For example, it has been pointed out that the media may play both a 'mystificatory' and 'de-mystificatory' role in relation to modern medicine (Karf, 1988). In addition, the food industry employs vast sums of money and uses the very best in sophisticated and expensive public relations efforts to influence media and channels of communication (see, for example, Balanya et al, 2000). As many of the examples in the chapters that follow show, it is the skilful use of the media by food companies that has ensured market success or increased sales for functional foods, not least in communicating messages about disease risk reduction.

Massive communications resources are being deployed in the promotional activities behind explaining the health benefits (or risk reductions) of foods and food components. From the mid-1990s onwards, much of the perception about food, health and risk in relation to functional food has been generated largely by food industry sources. To our knowledge there are no public health campaigns, similar to those developed as part of public health policy during the healthy-eating revolution, promoting the virtues of functional foods and their role in disease risk prevention.

The functional foods revolution presents the possibility that experts, often backed by competing business interests, may be seen to be guilty of distorting public perceptions when it comes to the presentation of risk in their functional food offering.

Distinctions between Absolute Risk and Relative Risk

The source of possible confusion over risk communication in functional food is the presentation of what is termed absolute as against relative risk. We should make it clear that both are important measurements in their own right, and it is not simply a case of 'either/or'. They are also calculations based on the same data sets, and are not terms derived from separate studies. They both describe for want of better words, the risk of an event occurring. Both are common terms in medical and pharmaceutical circles, but even here the distinctions and their implications are not always made explicit.

For example, pharmaceutical companies are keen to demonstrate results of drug efficacy in terms of relative risk since these sound far more impressive than if presented as absolute risk. The difference is in the way the mathematics is done, but the implications of the interpretation have immense consequences, as we illustrate below in relation to heart disease risk factors.

The easiest way to describe the distinction between absolute and relative risk is, in simplified terms, to describe how they might be used in the results from a drug study.

In the results of clinical trials on the use of a new drug, for example, estimates are made on risk and risk reduction. Many of these tend to talk about relative risk reduction and less about absolute risk reduction. The reason is simple. Using relative risk reduction it is easy to get results that demonstrate that risk reduction is very strong and, therefore, to make valid claims, for the sake of an example, that there is a 50 per cent reduction in the relative risk of a disease to patients treated with a particular drug.

Such figures can legitimately be calculated, because as the name implies, relative risk is calculated against something, or is relative to something, which in drug trials is the control group – it is, therefore, a measure of an 'odds ratio'. Thus, in a basic calculation of relative risk, if the risk of an event is 10 per cent in the control group, but is reduced to five per cent in the experimental group that is getting the drug or treatment, then the relative risk reduction is 10 minus 5, divided by 10 (that is, the ratio of risk reduction against, or relative to, the control group) – giving a relative risk reduction (RRR) of 50 per cent, which sounds very impressive! It is for this reason that RRR is used extensively in drug treatment advertising.

However, using the same figures, but using them to calculate absolute risk reduction, a different picture of risk emerges. There is a standard and accepted formula for absolute risk reduction in medical science. From this is derived the 'number needed to treat' formula. To calculate the 'number needed to treat' the formula is: 1 divided by the Absolute Risk Reduction number (ARR). So in our example, to calculate ARR would be $1 \div (10 - 5) = 5$ per cent, that is ARR would be 10 (the risk of an event in the control group), minus 5 (the risk of an event in the experiment group), which equals five per cent – a far less impressive figure than the 50 per cent RRR!

But the implications of ARR are even more striking. If, in this example, the ARR is five per cent, 20 people would need to be treated to prevent one ill-health event (that is, one divided by 0.05 – five per cent – derived from the 'number needed to treat formula'). For drug interventions, time periods are also crucial, and often require drug interventions for a number of years, but let us generously assume just two years in our example. Based on the ARR of our example, drug treatments would need to be given for two years to 20 people to save only 1. In many instances of drug therapy many years of intervention are needed to obtain results based on ARR calculations. Continuing our example, a common error in assuming RRR of 50 per cent is that 10 out of the 20 people would be 'saved'.

Both relative and absolute risk reduction calculations have their role to play in medical and scientific research – it is still an impressive result to demonstrate that an intervention can reduce relative risk by 50 per cent. Relative risk is also a powerful tool in calculating the likelihood of disease and illness, for example, in the incidence of cancer between smokers and non-smokers (calculated as the 'odds ratio' between smokers with and smokers without disease against non-smokers with and without the same disease). The result of this relative risk calculation is 'to avoid the risk of cancer, don't smoke'.

As Forde (1998) says, a reported 100 per cent increase or doubled risk for a disease is much more likely to bring an effect than a similar change reported as an absolute risk from one to two in 100,000, or as probability from 0.99999 to 0.99998. But 'number needed to treat' or ARR calculations are becoming so important, because of their implications for reducing the costs of drug therapy in health care services.

The discussion of relative and absolute risk offers both caution and opportunity for the food industry in respect to functional food. On the one hand there is the potential of misrepresenting or communicating poorly the true nature of the risk of disease, or the prevention of disease, to the general public in the promotion of functional foods that are represented as everyday food or beverage products. It could be construed as something of a sleight of hand to suggest that risk of disease or the prevention of disease applies equally across populations or population groups, especially if the risk reduction is seen as confined to a narrow band of foodstuffs. The danger of misleading the general public becomes more marked in calculations of risk viewed in terms of 'number needed to treat' (rather than relative risk) and when time periods over which risks might occur, are also included.

From this perspective, the assessment of risk poses considerable problems in reaching agreement in the development of one particular form of risk communication, namely health claims for food labelling and marketing. Risk communication through health claims used to promote foods and beverages to wide sections of the general public might give an unrealistic sense of security to many people and confusing messages about their total diet and its relationship to health. On the other hand, identifying and understanding risk for certain at-risk groups or populations offers great potential for functional food, especially as the genetic base of much disease and illness becomes better understood. In this case products could be narrowly targeted and deliver strong health messages to selected at-risk populations.

An Example of Decision Making Based on Risk Assessment

To illustrate the distinction between relative and absolute risk further we can give an example of how these problems confront physicians and the medical profession – and in a way which is very relevant to functional food. How risk thinking can influence decision making for physicians is illustrated by an example described in an article trying to guide doctors in the interpretation of medical literature (Oxman et al, 1994). The example involves a man aged 55 who, having his cholesterol level measured at a shopping mall, discovers that he has a high cholesterol level. He then visits his doctor for advice. The man does not smoke, is not obese, and does not have hypertension, diabetes mellitus, or any first-order relatives with premature coronary heart disease, but his cholesterol level is 7.9 mmol/l (the 'safe' cholesterol level is 5 mmol/l). Before recommending treatment, the doctor decides to find out just how large a reduction in the risk of heart disease this patient could expect from a cholesterol-lowering diet or drug therapy.

The resolution of this particular example raises important questions about functional foods aimed at the general population based on the notion of risk reduction. Oxman et al point out that the benefit that can be expected from interventions to lower cholesterol depends on the baseline risk of death from coronary heart disease and, possibly, whether dietary or drug intervention is being considered – the higher the risk of dying from the disease the greater the likelihood of benefit. Surprisingly the man in this example has a risk of dying of coronary heart disease of approximately one per cent over a period of ten years, despite his elevated cholesterol. Translating the risk to everyday practicalities, the authors say: 'You would have to treat approximately 1000 such patients for ten years with a dietary intervention to save a life.'

For such patients treated with drug therapy, it is also not certain that total mortality would be reduced. The results of this medical literature search are that:

> *'Drug therapy should be restricted to those at high risk, such as individuals with known coronary artery disease; whether diet therapy is worthwhile in low risk individuals (such as the patient in the scenario) is uncertain.'*

This example does help to illustrate the risk challenge of functional food. Is it now more appropriate for doctors to recommend functional food solutions (that is, dietary interventions or therapy), for low risk patients, high risk patients, both or neither, other than advising them to consume a diet based on healthy-eating dietary advice? In this example we have no idea what the answer might be. In short, will the disease risk equation be fundamentally changed now that functional foods are available?

The role of risk factors in the prevention of heart disease

Recent medical guidelines for the prevention of coronary heart disease suggest the risk equation is changing in a way that has profound implications for the way some functional foods are now being marketed. Ten years ago clinical recommendations on preventing cardiovascular disease focused primarily on managing individual risk factors, particularly raised blood pressure and cholesterol concentrations. Typically, separate guidelines were developed for each risk factor and treatment was recommended when that factor was above a specified level (Jackson, 2000). But as Jackson says: 'Over the past decade we have witnessed a remarkable change from these recommendations based on relative risk to ones based on absolute risk – that is, incidence.'

In practice this means concentrating on the relationship between a number of risk factors for heart disease rather than the emphasis on single risk factors as in the past. Important risk factors for heart disease are blood pressure, smoking and lipid concentrations (cholesterol levels). Other risk factors include age, sex, family history and patients with diabetes mellitus.

Nowadays European societies of cardiology, atherosclerosis and hypertension recommend that physicians should treat on the basis of absolute, not relative, risk of heart disease and should concentrate on those at highest risk. Taking the UK as a specific example, the British Cardiac Society, the British Hyperlipidaemia Association, the British Hypertension Society and the Joint British Recommendations on the Prevention of Coronary Heart Disease in Clinical Practice recommend that priority for treatment should be given to patients at high absolute risk of coronary heart disease, defined as the probability of developing coronary heart disease over a specified period. This approach uses a combination of risk factors as the basis for calculating absolute risk based upon well-established equations that are able to, with a high degree of statistical accuracy, predict incidence of disease (using the Framingham risk equations, Robson et al, 2000). Guidelines based on absolute risk ensure that undue emphasis is not being placed on an individual risk factor. In short, as Jackson writes: 'Using absolute risk to inform clinical decision making around cardiovascular disease prevention is no longer seriously questioned.'

The editorial in the same issue of the *British Medical Journal* (BMJ) in which Jackson writes, puts it more bluntly:

> *'...with the wisdom of hindsight much of our response looks unintelligent: treating people, regardless of their absolute risk, when they cross a particular magic line of blood pressure or lipid concentration.'*

Therefore, just as the food industry has come up with its biggest food and health idea – designing foods for single risk factors, notably the single risk factor of elevated cholesterol in respect to coronary heart disease – the medical profession is fast moving on to the concept of absolute risk which addresses combinations of risk factors. Such new thinking in the medical profession, suggests the overemphasis by food companies on single-risk-factor functional foods might become an idea 'dead on arrival'. For food marketers it will become more important than ever to position their functional foods as one component in the total management of an individual's health and well-being. One real world example is Unilever's marketing of its Flora brand margarine (on a cholesterol-lowering proposition) under the umbrella of The Flora Project, a nutrition marketing approach which addresses all the risk factors for heart disease and which we cover in more depth in Chapter 10.

It is worth considering the origins of this apparent U-turn in medical thinking. But first we should point out that we are not trying to understate the seriousness of cardiovascular diseases as the major cause of death in many developed countries. They still represent a major burden on health care services and we are not in any way implying that coronary heart disease in Western society is not an important health care problem. For example, between 1981 and 1991 in Britain, there was a 31 per cent fall in the prevalence of heart attacks in both men and women. But out of a population of around 57 million, 1.4

million people have had a heart attack in the UK (about 300,000 heart attacks each year); and there are two million sufferers from angina (Unilever Research Laboratorium, 1998). Elevated cholesterol in populations has long been established as one of the classic risk factors in coronary heart disease and remains important.

However, in a historical perspective, it can be seen why there has been so much attention paid to single risk factors like elevated cholesterol. From the 1950s onwards, epidemiological studies, led by Framingham in the US (Dawber, 1980), identified personal characteristics as risk factors for the development of coronary heart disease in previously healthy people. Of these modifiable factors, distinct from age and sex, cigarette smoking, blood pressure, and total blood cholesterol were most consistently and powerfully implicated. These three became the classic risk factors for heart disease (Kuulasmaa et al, 2000).

Later, in the 'Seven Countries' study (Keys, 1980), the investigators showed these risk factors to be important within populations in different countries, but claimed that the key determinant of population differences in risk was the mean concentration of blood cholesterol, which was in turn related to dietary saturated fat (supplied, mainly in the Western diet from animal sources such as dairy products, meat products and through the use of animal fat in processed foods).

Studies across geographically diverse populations have been less successful in explaining differences in coronary heart disease mortality, hence the interest in ideas such as the French paradox and the Mediterranean diet. However, few other risk factors have approached the power, consistency or prevalence of the original three (Kuulasmaa et al, 2000). Hence, for historical reasons there has been a strong emphasis on the single risk factor of cholesterol. New medical thinking, while not diminishing this as a classical risk factor, is once again re-emphasizing the combination of classic risk factors in combination with other risk factors as a basis for interventions and treatment.

Functional food, the medical profession and risk

The goal of many food companies is to find publications in the medical literature supporting or demonstrating disease risk reduction for dietary components relating to their products or ingredients. Single studies published in medical journals are being used as the basis for some functional food marketing. But these need to be seen in context. Today's medical literature is daunting in its volume: 7500 articles a week are added to the medical profession's premier

Box 4.1 Case Study: Lycopene and Prostate Cancer Risk

The importance of the concept of risk in relation to functional food can be illustrated by the example of the consumption of the dietary component lycopene and its possible role in the reduction of the risk of prostate cancer in men. Prostate cancer is a leading cause of mortality in US males and lycopene is one of the major carotenoids in Western diets, its main source being in tomatoes (lycopene makes tomatoes red!). In recent years tomatoes, particularly processed tomatoes, have been heavily promoted on the basis of their lycopene content and their potential in risk reduction in relation to prostate cancer. This is based on interesting, but far from complete, peer-reviewed, published scientific research. For example, according to a review of the scientific literature on lycopene in relation to human health by Gerster (1997) 'remarkable inverse relationships between lycopene intake or serum values and risk have been observed, in particular for cancers of the prostate, pancreas and to a certain extent of the stomach'.

However, the promotion of lycopene in relation to risk reduction of prostate cancer has rested primarily on two prospective studies that evaluated the relationship between tomato consumption and the risk of prostate cancer. One study was conducted on a cohort of Seventh Day Adventist men (Mills et al, 1989) and, more recently, calculations derived from the health professionals follow-up study in the US (Giovannucci et al, 1995). In the last study a relative risk reduction of 35 per cent was observed on those men developing prostate cancer, based on a consumption frequency of greater than ten servings of tomato products per week versus less than 1.5 serving per week.

In his review of lycopene and human health, Clinton (1998) cautions:

> *'...if lycopene indeed contributes to lower risk of prostate cancer, the mechanisms remain speculative...overall, the hypothesis that the consumption of tomato products reduces prostate cancer risk via lycopene warrants further research, but much more evidence is necessary before a causal relationship can be established.'*

Rao and Agarwal (1999), confirm Clinton's point of view that current evidence remains 'suggestive' rather than proof of a causal relationship, and also remind us that human intervention trials have not been performed investigating the effectiveness of lycopene intake on lowering the risk of chronic disease.

The successful promotion of lycopene in relation to prostate cancer, could suggest to many consumers that a lack of lycopene in the diet might be the only risk. In Chapter 10, for example, we describe in a case study the multimillion dollar promotion of lycopene's health benefits by the tomato products' industry. While we support the promotion of good dietary practice, the point we are drawing attention to here is that the case of lycopene and prostate cancer also highlights the complexity of understanding the nature of risk and disease. As detailed earlier, there is suggestive evidence about the importance of lycopene in the diet, but the relative consumption or otherwise of lycopene in relation to prostate cancer risk is just one of a large number of potential risk factors associated with developing prostate cancer, many of these as yet unidentified. Put in context, nutrition accounts for only five per cent of the possible risk factors in developing prostate cancer (McKinlay and Marceau, 2000).

McKinlay and Marceau write that some 60 risk factors for prostate cancer have been identified in the professional literature. But they point out that, unfortunately, the data on a particular risk factor available in one study are often not available in other data sets reporting on other risk factors. Consequently, it is impossible to estimate precisely the relative weight of particular factors and their combined contribution to the explanation of disease. They write:

> *'Our on-going Massachusetts Male Ageing Study has data not on all 60 risk factors for prostate cancer, but on 36 of them...after many decades of risk factorology, more than two-thirds of the contributors to (causes of) prostate cancer remain unidentified. Similar situations exist for other major diseases, such as coronary heart disease, diabetes and stroke.'*

A massive 82 per cent of the contributory risk factors for prostate cancer are unknown. Out of the 18 per cent of known risk factors, nutrition contributes a mere five per cent – age is almost as relevant a risk factor as nutrition.

database called MEDLINE, which now holds 11 million publications built up between 1966 and 2000. This volume of information is fuelled by the 4200 medical journals currently in publication, and as the volume of literature has grown, risk has become one of the trendiest subjects, causing what one researcher describes as a 'risk epidemic' to be rooted in scientific medical journals (Skolbekken, 1995).

Skolbekken found an explosive increase in the number of articles with the term 'risk' in the title or the abstract in the US, the UK and Scandinavia between 1967 and 1991. The increase ranged from 20- to 80-fold in the prestigious *BMJ*, the *Lancet* and the *New England Journal of Medicine*. In epidemiological journals studied, 50 per cent of articles in the last five years reviewed contained the term 'risk'.

However, the medical profession is far from being in the vanguard of thinking about risk and would stand as an interesting case study in the social construction of risk. In fact medical care may be one of the last sectors of society in industrial countries to embrace fully probabilistic thinking (Fox, 1999).[3]

Marketing of functional foods through physicians

Many food companies believe it is crucial to include health professionals as part of their marketing strategy, particularly to gain the support of physicians, but there is as yet little concrete empirical evidence to suggest why this is a good strategy for food. There is some evidence that the extensive and expensive marketing and advertising efforts of pharmaceutical companies influences drug prescribing patterns of medical doctors, so it will be interesting to see if food company marketing efforts can improve the acceptance of functional foods.

While acknowledging that the medical profession has a wide role in health care, including health promotion, in general terms physicians deal in medicine and not food. The implication of this for functional food is that, by definition, physicians see a particularly self-selecting population with a range of conditions, not all of which may need dietary advice. So the apparent influence of the medical profession is possibly overestimated. For example, in a consumer survey by the American Dietetics Association (ADA), while physicians are rated highly as a valuable source of information about nutrition by 92 per cent of those surveyed, only 11 per cent had used physicians as a source of information (ADA, 2000).

When patients go to see the physician, it will probably have nothing to do with nutrition. For example, Forde (1998) points out that while health care consumption has risen dramatically for minor conditions, in 30 to 60 per cent of the visits to primary care physicians no medical condition was found to account for the patients' symptoms. In the US, the National Mental Health Association (NMHA) reported in 1997 that 75 to 90 per cent of all visits to physicians are stress-related (*Fast Company*, March 2000). Job stress is estimated to cost US industry US$200 to $300 billion annually in absenteeism, diminished productivity, workers' compensation, and direct medical, legal and insurance fees (Fischer, 1998).

In addition, nutritional risk can involve some unexpected elements not necessarily within the remit of the medical profession. For example, low socio-economic status remains the single most potent determinant of morbidity and mortality – and the only thing wrong with poor people is that they don't have enough money, which is a different sort of curable condition. McIntosh et al (1990), in a study of nutritional risk and the elderly, found nutritional risk is lower among those who have a greater number of companions, who eat with others, and who receive both help with cooking as well as shopping for groceries. Those with religious networks were at even lower risk. But one of most important indicators of those at nutritional risk was the wearing of dentures!

The food marketing efforts directed to physicians, however, may have some unexpected benefits. For example, research has shown that, generally speaking, physicians do not always have a deep knowledge of nutrition perhaps as the result of inadequate training in this area (Heimburger et al, 1998; Temple, 1999). For reasons we have already outlined, to associate medicine with functional food, particularly in relation to disease risk reduction, may need considerable reassessment in future years.

Risk and biomarkers or endpoints

Above all else for the future success of the functional foods revolution – seen from the scientific perspective – is the question of risk in relation to biomarkers or endpoints (also known as surrogates). Biomarkers are chosen because they are thought to be on the causal chain leading to a clinical outcome (that is, fewer myocardial infractions, strokes or deaths). In food marketing, one of the most widely used biomarkers currently targeted by a growing number of products is blood cholesterol levels, a well-established marker in the risk of heart disease. An objective of the functional foods revolution is to identify and use a whole range of biomarkers in product development.

In research on functional food, the development and application of biomarkers is extremely important, but a key issue is the emphasis put on the importance of the effects of food components on 'well-identified and well-characterized target functions in the body that are relevant to health issues, rather than solely on reduction of disease risk' (Diplock et al, 1998, in the consensus document produced for the EU Functional Food Science in Europe programme, coordinated by ILSI Europe). In the consensus document the authors set out a list of potential target functions for functional foods, for example, maintenance of desirable body weight; and possible markers for these. This approach opens up a new pathway of risk and disease reduction in relation to food components and the way food relates to health.

In an excellent earlier article by Diplock et al (1998) the concept of biomakers in relation to antioxidants is reviewed. The authors write that epidemiology studies are necessary to quantify the impact of antioxidants on disease aetiology and intervention trials to formally test the efficacy of enhancing intake of antioxidants. In evaluating these health benefits, the authors suggest that 'hard endpoints' (disease incidence, or recurrence and mortality) should be used. Alternatively, they point out, intermediate endpoints may be effective, provided they are genuinely predictors of the disease of interest. Importantly they say: 'Biomarkers of exposure should accurately reflect relevant dietary intake or body status and early disease markers should have predictive value for the hard end-point.'

For the functional food science programme in Europe, the concept of biomarkers was seen as crucial to develop universal health benefits for functional food. These in turn can be used to establish health claims for food components.

Again the concept of biomarkers and disease reduction becomes a very technical and complex scientific debate, compounded by the nature of the diseases and the food components being investigated. The use of endpoints is a well-established method in drug research, but some commentators believe there are major shortcomings that have to be overcome if endpoints are to be useful. Such a critique hinges on the understanding of risk reduction and the causal relationship between markers and disease. Such issues will need to be addressed if functional foods are, after appropriate risk/benefit analysis, to be established as universal markers for health and well-being.

For drugs, data on surrogate markers (endpoints) such as blood pressure or body weight have often been used to support approval, for example, of new pharmacological treatments for cardiovascular risk factors (Psaty et al, 1999). Using surrogate endpoints in drug research has a number of advantages. For example, clinical trials can be a lot shorter (weeks rather than years), patient samples can be smaller, and the subsequent costs are reduced. However, for some commentators, the use of surrogate endpoints in the assessment of cardiovascular disease drugs, needs to be viewed with caution (see Temple, 1999). Temple raises major concerns about the reliability of surrogates when used in drug development. One of these concerns, and pertinent to functional food science, is that the surrogate may not be valid. In other words, a candidate surrogate endpoint may not in fact be causally related to the clinical event.

He also points out, again in the case of drugs, that whether a surrogate or intermediate endpoint is appropriate to use as a basis for marketing is a matter of scientific judgement. In developing health claims for foods based on endpoints there must be some mechanism (and transparency) for defining how the scientific judgement has been reached. Temple concludes in his assessment by saying: 'Surrogate endpoints are thus neither consistent successes nor consistent failures.' In other words, the value of surrogate endpoints for drug assessment is fraught with a level of ambiguity. If a similar ambiguity is found for biomarkers in relation to dietary components, it will most surely add to the problem of reaching agreement for health claims based upon biomarkers which are appropriate for the general population.

Clydesdale (1997) reminds us of the importance of health claims for food and how these can influence consumer behaviour and potentially affect public health. In one of the best papers on health claims we have seen, Clydesdale makes a compelling case for functional foods and health claims and their potential for public health based upon internationally agreed scientific criteria, including the use of surrogate markers or biomarkers. But to implement Clydesdale's proposals, in our view, this particular debate would need to be widened and become more inclusive if policy makers and the industry are to abide by international decisions.

Conclusion

At the beginning of this chapter we said our purpose is not to write a book within a book on risk, but simply to raise the issue as one not adequately addressed in wider discussions on functional foods and their introduction into food supply. In particular, the communication of risk through, for example, health claims on product packaging, should be researched in greater depth to inform policy making. In the EU, for example, there is the possibility that health claims for food labelling are going to be introduced without empirical evidence to their effectiveness or how they should be communicated.

The discussion on risk, food and health is very much a specialist one. But it is important to nutrition marketing and functional food if the food industry is not to be accused of misrepresenting scientific data and medical research in the pursuit of market share and profits through functional food. Perhaps the biggest lesson from the communications of risk in relation to disease reduction for food is that there is much to be said for keeping it simple – even in functional food. For example, people know that when their life is at stake, regardless of the risk, they end up 100 per cent alive or 100 per cent dead! (Goodman, 1999).

Chapter 5

Health Claims and Functional Food

One of the most taxing and time-consuming aspects of public policy in relation to functional food is health claims. For many, the international development of functional food hinges on the ability to make health claims for specific products. Some companies cite the lack of ability to make health claims as a major reason for their reluctance to commit significant funds to rigorous human testing of efficacy and toxicity of products (Stephen, 1998).

Put simply, an ideal policy on health claims will allow food producers to tell consumers, preferably in a strong and unambiguous fashion, the nature of the health benefit they may derive from consuming a particular food or beverage. By implication, the health claim will persuade consumers that a particular product is preferable to consuming competing or similar products which don't carry health claims. Currently, health claims are illegal in food labelling and marketing (except in special circumstances – see Chapters 6 and 7 for examples) and, it is reasoned, that this is holding back the market for functional food.

There is a compelling case for allowing health claims: how else are consumers going to be able to differentiate between products that confer scientifically validated health benefits and make legally evaluated claims, against those that are unproven, misleading or even fraudulent? In addition to such consumer protection, Professor David Richardson captures the core difficulty facing food manufacturers and their frustration over limitations on the use of health claims when he writes:

> *'...what could be extremely frustrating to reputable food manufacturers is that functional foods backed by considerable research efforts and investment could be undermined by, and appear to be the same as, those that are crude, carelessly made, lacking substantiating evidence, and, at worse, fraudulent. If the consumer did not believe or trust the ability of a product to provide the stated benefits, the long-term credibility of the industry would soon be damaged.'* (Richardson, 1998)

With many food regulations there is limited or no scope for making a claim on a food or drink product that implies or suggests, directly or by implication, that it might be capable of preventing, treating or curing human disease. In wanting to make health claims, functional foods, already handicapped by lacking any legal definition, often fall into a legal grey area between food and drug. This is the crux of the functional food development crisis and the source of the complexity and controversy over its place in food supply and potential benefit to individual or public health.

In this chapter we explore the consumer market and policy pressures around the issue of health claims to highlight the complexities already alluded to. We make a detailed case study of the market, consumer and policy background that led to the setting up of the Joint Health Claims Initiative in the UK. The UK is generally regarded as a hostile market for functional foods due to widespread consumer advocacy concern about such products. The Joint Health Claims Initiative is a unique collaborative effort by the food industry, consumer advocacy representatives and food law enforcement agencies to address the issue of health claims and functional food through the development of a health claims code of practice.

The UK case study is an opportunity to examine the wider issues and difficulties in setting up a secure framework for making health claims – many of the public policy issues are the same as other EU member states. Further, in Japan and the US, developments in the market for functional foods cannot be understood without detailing attempts to facilitate the use of health claims for foods and beverages and these are covered in detail in Chapters 6 and 7. A substantial part of the discussion in these chapters is how regulators, policy makers and industry have wrestled with each other over this. The importance of the issue of health claims is further enhanced by international activity, and by the food industry in particular, to establish 'universal' health claims to facilitate global competition in foodstuffs.

The Major Problems in Regulating Health Claims

The Codex Alimentarius Commission, the international food standards agency, defines health claims as 'any representation that states, suggests or implies that a relationship exists between a food or a nutrient or other substances contained in a food and a disease or health-related condition'. A health claim is, therefore, fundamentally different in principle from a nutrition claim, a nutrient function claim or nutrient content claim.

Central to the development of health claims for products and ingredients is to show both safety and efficacy (that is, do they actually do what they claim?). While safety is always a central issue in food supply, functional food developments are introducing new questions around safety. For example, what are safe upper limits for certain nutrients? Even humble vitamins and minerals, especially in large doses, are subject to this safety concern as their nutraceutical properties are more clearly delineated (Richardson, 1998).

Also of concern, on grounds of safety, is the wider use in the food supply of other substances, including natural components such as herbs, plant sterols and certain bacteria, consumed in quantities or in ways not normally established within traditional dietary or consumption patterns. Finally, there are safety concerns about novel foods and ingredients and to what extent many functional foods and ingredients should be considered novel and be subject to novel foods regulation. A major area of potential difficulty here would be the application of biotechnology to produce novel functional foods or ingredients.

A further central issue to the use and development of health claims is whether they should be allowed as generic claims or as product-specific claims. For example, Benecol has demonstrated scientific validation for the cholesterol-lowering properties of its margarine-type spread as consumed. This is product-specific and could not be extrapolated in other forms to other products. A generic claim can apply to any food substance that has a demonstrated health benefit. For example, in the US, generic health claims have been established for a number of diet and health associations such as 'while many factors affect heart disease, diets low in saturated fat and cholesterol may reduce the risk of this disease' (see Chapter 7). Generic claims carry no exclusivity and can be used by any food producer meeting the appropriate nutritional criteria established to allow the health claim.

Voluntary Health Claim Guidelines in other EU Countries

Sweden was the first European country to develop a programme for the use of generic health claims in the labelling and marketing of food products. The programme first came into effect in August 1990, but was revised in August 1996. The code allows eight connections between diet-related diseases, or risk factors – obesity; cholesterol level in the blood; blood pressure; atherosclerosis; constipation; osteoporosis; caries; and iron deficiency. However, a survey of the

Swedish market in 1998 found that these generic claims were not in widespread use (Heasman and Mellentin, 1998).

Several other European countries, in addition to the UK and Sweden, have also taken initiatives to establish voluntary codes for the use of health claims.

In The Netherlands, the Dutch Nutrition Centre (Voedingscentrum) has drawn up a code of practice entitled 'Assessing the Scientific Evidence for Health Benefits stated in Health Claims on Food and Drink Products 1998'. The code is supported by numerous industry and consumer organizations, and the regulatory authorities. The code lays down a procedure for assessing the scientific evidence for claimed health benefits. For example, Unilever's cholesterol-lowering margarine with plant sterols has gone through the stated procedure and a cholesterol-reduction claim has been approved.

A code of conduct in relation to health claims has also been drawn up in Belgium, where the Federation of Agriculture and Food Industries (FIAA) has been the driving force.

In Spain, the Spanish Food and Drink Federation has signed an agreement with the Spanish Ministry of Health on the issue of health claims. The agreement is intended to clarify the legislation applicable to the advertising and labelling of foodstuffs and their properties in relation to health. Health claims – as defined in the agreement – are not prohibited and may therefore be used provided certain conditions are met.

In Finland, the National Food Administration has published guidelines for permissible health claims, so-called vital function claims. These guidelines are currently under reconsideration and indications are that Finland will adopt a more liberal attitude. However, the authorities are concerned with the fact that legislation should not infringe upon current legislation, including the EU labelling directive.

Furthermore, Germany, France and Denmark are considering legislation in this area, or more precisely these countries are considering how current national legislation regarding health claims should be interpreted in the future, given international and scientific developments. Denmark is also drawing up a list of generic health claims along the lines of the US and Swedish models.

Regulatory Developments in Australia and New Zealand

Australia and New Zealand are carrying out a unique pilot study to evaluate health claims on food labelling. The Australian and New

Zealand Food Authority (ANZFA) set up the health claim trial at the end of 1998, based on the well-documented scientific link between intakes of folic acid and the reduction in the risk of neural tube defects in newborn infants.

Companies participating in the pilot study, such as Kellogg and Sanitarium, are able to carry a direct health claim for folate-fortified foods stating that if women consume at least 400 micrograms of folate a day then the risk of neural tube defects in pregnancy, which can lead to babies being born with spina bifida, is greatly reduced. The 'Folate – make it part of your day' campaign was launched in February 1999 in Melbourne, Australia. Manufacturers are using folate-related benefits as part of their overall advertising strategies and supermarkets are displaying in-store information.

ANZFA is keen to stress that eating a combination of folate-rich foods is vital, rather than taking a supplement or eating fortified cereals or breads. More than 100 foods have been approved to carry the health claim including eggs, legumes, various fruits and vegetables, together with a wide range of bread, cereals and juices. Processed foods making the claim also have to meet criteria for fat, salt and sugar. Part of the trial evaluation is to assess whether health claims make any difference to sales of products using the claim, in addition to the main public health goal of increasing folate intakes.

The folate health claim trial resulted from years of consultation on functional food in Australia. This started in the early 1990s and culminated in a concept paper on health and related claims, published in February 1996 after a three-year consultation period. The consultation exercise included a number of public meetings and was inclusive of all interests. The consultation revealed just how contentious this topic was in Australia. Many public health professionals and consumer agencies opposed the concept of health claims and functional food, regarding these as motivated by marketing objectives and the inappropriate interpretation of medical research by scientists who are not experts in public health. The food industry, on the other hand, argued there was the opportunity to improve the health of significant sectors of the population through such food. Although a little dated, the Australian concept paper, in our view, is still one of the best policy papers on functional food and succinctly captures the key issues.

Science Is Needed to Demonstrate a Health Claim

Demonstrating efficacy – whether the science behind a product or ingredient can justify a health claim – is fraught with difficulties, not

Box 5.1 Basis for the Scientific Dossier to Substantiate Health Claims

Dietary significance; intake; extent of use
Interactions with other components of diet; bioavailability
Presence of anti-nutritional factors
Quantitative effects; dose response
Impact on metabolic pathways and physiological function in humans
Nutritional composition
Overall toxicological assessments including allergy/intolerance factors
Potential effects on vulnerable groups: for example, the young and elderly.
History of safe use; previous human exposure
Storage, preparation and instructions for use
Direct effects on patho-physiological processes
Relation to current dietary recommendations/targets
Technical details of processing and product specification

Source: Adapted from Richardson, 1998

least the question of how much science is enough to be accepted as demonstrating a particular health claim (the 'significant scientific agreement debate') (see Chapter 4).

From a public health point of view Lawrence and Rayner (1998) come to the conclusion that there is the need to maintain a general prohibition on health claims, but acknowledge that specific exemptions could be made supported by scientific substantiation.

Richardson (1998) again emphasizes that the substantiation of health claims must be based on sound science which includes the synthesis of the existing literature, epidemiological, metabolic, animal studies, human clinical evaluations and mechanistic data. He suggests a guide to manufacturers on how to prepare a comprehensive scientific dossier to support health claims. (See Box 5.1.)

Lawrence and Rayner (1998) also detail the range and types of evidence that, in their view, should be required to support a health claim application. In addition, they provide a useful conceptual framework of where health claims lie in relation to other information on food labels. They say the intention of their framework is to illustrate the relationship between the different types of potential information that may appear on food labels and to suggest where claims for functional food may fit within the health claim context.

The Debate on 'Sound Science' is Far From Over

The one area of complete agreement is that health claims should be based on sound science. What is not so certain is what constitutes this. Rayner (1998) proposes a 'systematic review' as a method for assessing the validity of health claims. Systematic review has become an established method for assessing the evidence for the effectiveness of medical interventions and Rayner believes such methods could prove effective in establishing the scientific validity of health claims in a more robust manner than in current methods, such as 'expert consensus'.

Two of the cornerstones of the systematic review are making sure all the evidence is considered and accounting for publication bias. Rayner gives a pertinent example to illustrate this latter point, writing that publication bias in a diet-related area has been clearly demonstrated in a meta-analysis of the effect of garlic in the treatment of moderate hyperlipidaemia (that is, elevated levels of blood cholesterol). Statistical analysis on the published studies of the effect of garlic on blood cholesterol shows that it is likely that many small studies showing a negative effect of garlic have not been published. This evidence of publication bias now means that the published evidence cannot itself resolve whether there is an effect or not (Rayner, 1998).

Lawrence and Rayner (1998) suggest stakeholders have tended to initiate and frame the public policy debate on health claims around opinions and speculation, rather than empirical evidence. We have drawn attention to the lack of academic research for the proposed development of health claims (Heasman and Mellentin, 1999).

One of the most graphic examples of the 'scientists' developing policy on communications and health claims utilizing virtually no research evidence from the social science and policy literatures, is the European Union Functional Food Science programme, project-managed by ILSI Europe. While producing outstanding reviews of the science in relation to functional foods (see Bellisle et al, 1998), in the programme's final consensus document (Diplock et al, 1999), the authors conclude by mapping out a policy and communications strategy for health claims and functional foods. In this section of the consensus document none of the policy or academic research in the area of communications and public health practice is referred to (see for example, as an obvious starting point, Maibach and Parrott, 1995).

In view of the emphasis throughout the programme on sound science and research this appears as an astonishing omission. There is now a growing literature in public policy, social marketing and other academic fields on the communication of health issues to the public, its effectiveness and strategies. This work could inform the development of policy and strategy for the communication of the health benefits of functional food, in particular the communication of science, which Diplock et al rightly say is 'an inherent part of functional foods'. It is a mystery why work such as this has not been drawn upon by the scientific community, who feel free to pronounce on health claims communications policy as though empirical and theoretical work in these areas is non-existent.

The good news is that the debate about health claims in relation to functional food is still evolving and can become more academically inclusive, but time is pressing to get it right. For example, the EU is expected to have legislation in place allowing health claims for the labelling and marketing of foods around July 2002 (Affertsholt, 2000). What is remarkable is the international convergence of public policy to accommodate health claims for functional food. In the world of food legislation, not renowned for speed of change, in the space of a few years nations have sometimes been prepared to reconsider public policy and introduce new measures to enable health claims. This, if anything, shows the power of the functional food concept and the powerful advocates behind it. We turn to one such initiative by way of example, to investigate the development of health claims in more detail.

Case Study: Developing a Code of Practice for Health Claims in the UK

The UK, like other EU member states, is subject to EU-wide legislation, but the implementation and interpretation of this legislation is carried out on a national basis. So, for example, Council Directive 79/112/EEC on food labelling in the EU was incorporated into UK food law as the Food Labelling Regulations 1996. Since 1994 there has been considerable policy activity in the UK on functional food and, in many respects, a marked shift in policy thinking in this area. In particular, concerns about health claims for food and drink, and about their potential for abuse, have brought together consumer advocacy groups, the food industry, regulators and enforcement authorities to form a Joint Health Claims Initiative (JHCI) to devise a voluntary code of practice for health claims for foods. The JHCI did not happen in isolation, but came about through a process of lobbying and negotiation, mainly on the part of consumer organizations who were, at first, excluded from the functional food policy process.

Policy developments were promoted initially by both the Ministry of Agriculture Fisheries and Food (MAFF) and the food industry. Both saw functional food as offering UK companies the potential for competitive advantages in international food markets, but policy action has also been prompted by the critiques raised by consumer advocacy groups. Policy initiatives have also to be seen against a hostile consumer advocacy environment for functional food – which we describe below. The JHCI, in part, can be seen as a response to this consumer environment. It was set up in 1996 and by February 1998 had produced a Code of Practice on Health Claims on Foods for the UK. Below we examine this unusual event and the context in which it took place.

Current controls in the UK and EU

There is no legal definition in the UK for functional foods, but a working definition developed by MAFF was 'a food that has a component incorporated into it to give a specific medical or physiological benefit, other than a purely nutritional benefit'. They are common or normal foods with the same appearance as conventional foods, but with amended composition, that are still eaten as part of the usual diet. Interestingly, Cockbill, then Head of the Consumer Protection Division at MAFF, raised in an article the question of whether there is any need to attempt a definition for functional foods, legal or otherwise. He pointed out that functional foods are in the same situation as any other foodstuffs that are placed on the market and about which the manufacturer wishes to make a claim. He wrote:

> *'If the foodstuff can adequately fulfil the function for which the claim is made, is hygienically produced, and its labelling provides accurate information for the consumer, the marketing of that foodstuff should not run up against any difficulty. Moreover, it does seem unlikely that there is either room or need for any more particular rules covering functional foods.'* (Cockbill, 1994, p4)

As Cockbill noted, there are no specific controls on functional foods but, unless they are medicinal products within the meaning of the Medicines Act with medicinal product marketing authorization, functional foods are subject to the existing food laws. These are as follows:

> *'Under the Act it is an offence to render food injurious to health. It is also an offence to sell or keep for eventual sale, food that* inter alia, *has been rendered injurious to health, or is not of the nature, substance or quality demanded by the purchaser, or is falsely described or misleadingly presented, advertised or labelled.'*

The Food Labelling Regulations set out the labelling requirements which apply to most packaged foods and also cover certain claims made about a food, either on the label or in advertisements, thus extending the general provisions of the Food Safety Act of 1990 and Trade Descriptions Act on misleading descriptions. The general rule is that label information must be true and not misleading, and this applies equally to claims. In addition there are specific requirements in respect of certain claims:

- The 1996 Food Labelling Regulations prohibit an express or implied claim, in labelling or advertising, that a food is 'capable of preventing, treating or curing human disease' unless the food has a product licence issued under the Medicines Act. Disease is interpreted as 'any injury, ailment or adverse condition, whether of body or mind'. This effectively amounts to a ban on those health claims, which make references to human diseases;
- Claims that a food is suitable for a particular nutritional use (defined in EC Directive 89/398) must be accompanied by information about the food's composition justifying the claim, plus a nutrition panel listing certain macronutrients and, if necessary, the relevant micronutrients. This is the UK implementation of PARNUTS;[1]
- Claims about nutrient content may only be made if the labelling also carries a nutrition panel listing certain macronutrients and, if necessary, the relevant micronutrients;
- As far as vitamin and mineral claims are concerned these can only be made for 18 scheduled substances listed in the 1996 regulations and on condition that certain criteria are met.

There are no specific controls on health claims such as 'believed to promote healthy digestion', but they must be true and not misleading. The borderline between these and implied medicinal claims is not always clear. The government's Food Advisory Committee recommended a more systematic means of control in its 1990 report on food labelling. The government decided not to adopt it while there was a prospect of further European controls on claims.

If the functional food is a novel ingredient or contains a novel ingredient the manufacturing company will invariably request a safety clearance by submitting a dossier of evidence to the Advisory Committee on Novel Foods and Processes (ACNFP), even though this is voluntary.

MAFF's functional foods initiative

Despite these provisions and the apparent lack of any need to make functional foods a special case, MAFF, under the Agro-Food Quality LINK Programme, commissioned a Functional Foods Initiative in 1994. A conspicuous omission from this MAFF initiative was its complete failure to consult or include consumer groups. While the contractors were able to consult with more than 100 companies and academic researchers, they did not consult with any consumer or public interest organizations. Further, in a series of focus groups on functional foods held at the end of 1994 one actually addressed consumer issues. This focus group brought together three people from research organizations and five industry representatives. On the basis of the focus group the following recommendations were made:

- Functional foods should be considered as ordinary foods, without the need for special regulation or codes of practice. It was the general consensus that the regulatory framework currently in force controlling health claims was effective in preventing misleading information to the consumer (it should be noted that in other parts of this review industry representatives are shown to take an opposite view);
- On the basis that the food industry believes it has a role to play in educating the consumer on nutritional matters, MAFF may wish to consider whether it should allow claims based on the knowledge concerning the relationship between foods or food components to be made, eg folic acid and neural tube defects. The ability to do so is seen as fairly critical to the development of functional foods;
- MAFF should consider the comparative effectiveness of health claims versus compositional labelling in modifying dietary patterns, especially as the food industry appears to have serious reservations about the latter.

MAFF held a conference on 'Functional Foods: Status, Opportunities and Research in the UK' on 23 March 1995 as part of their initiative. In June 1995, only after representations from the National Food Alliance, an umbrella organization for a large number of consumer organizations with an interest in food, did MAFF agree to a series of meetings to consult with consumer representatives and groups in relation to functional food.

The outcome of MAFF's wider consultation process was a joint meeting between the different groups with interests in functional food and health claims on 10 November 1995. At the end of this process MAFF was still of the opinion that there did not appear to be a strong case for additional regulation for functional food and health claims. However, other parties involved, that is retailers, manufacturers, trading standards officers and public interest groups, did agree on the need for control of food claims and that in the absence of EU legislation, voluntary guidelines would be helpful and should be developed. This meeting led to the eventual birth of the Joint Health Claims Initiative.

Official review of functional food and health claims

Meanwhile in March 1996, the government's Food Advisory Committee (FAC) announced a review of health claims by issuing draft guidelines on health claims in foodstuffs. This was somewhat ironic since the FAC had already made recommendations on the regulation of health claims in its review of labelling issues in 1990, advice which MAFF had then rejected. MAFF sent out consultation letters to just under 200 parties on 16 December 1996, including manufacturers, retailers, and enforcement bodies and consumer organizations.

In its earlier 1990 review of labelling, the FAC set out principles that could be used to control permitted health claims. These were:

- The claim must relate to the food as eaten rather than to the generic properties of any of the ingredients;
- A food (when consumed in normal dietary quantities) must be able to fulfill the claim being made for it and adequate labelling information must be given to show consumers that the claim is justified;
- The label should give a full description of the food to ensure that selective claims, even if true, do not mislead and any claim should trigger full nutrition labelling;
- The role of the specific food should be explained in relation to the overall diet and other factors.

In the 1990 review, the FAC concluded that a prior approval system for claims would be unnecessarily burdensome. A system of label notification – where companies submit details of claims and the scientific evidence behind them to the authorities, but do not require approval before marketing – was also considered. The review concluded that it was 'on balance not justifiable'. The 1996 consultation exercise made it clear that developments since 1990 meant the options needed review.

The FAC draft guidelines on health claims on foodstuffs ran to some seven pages and were seen as a constructive intervention in the UK's health claims debate. For example, the FAC provided a useful and comprehensive definition of what a health claim is:

> *'Any statement, suggestion or implication in food labeling or advertising that a food is beneficial to health, but not including nutrient content claims (for example, that a food contains low fat, reduced cholesterol or high fibre) nor medicinal claims (that is, that a food is capable of curing, treating or preventing a human disease or any reference to such properties).'*

Further, the FAC said the term 'health claims' could be subdivided into:

- Claims which refer to possible disease risk factors (for example, 'can help lower blood cholesterol');
- Claims which refer to nutrient function (for example, 'calcium is needed to build strong bones and teeth');
- Claims which refer to recommended dietary practice (for example, 'eat more oily fish for a healthy lifestyle').

Human disease was defined to include any injury, ailment or adverse condition, whether of body or mind.

The guidelines say that positive messages that make links between food and health may help purchasers to maintain or move towards a healthy diet, and give examples of health claims of a nutrient–function type which command a general scientific consensus:

- 'Calcium aids in the development of strong bones and teeth';
- 'Protein helps build and repair body tissues';
- 'Iron is a factor in red blood cell formation';
- 'Vitamin E protects the fat in body tissues from oxidation';
- 'Folic acid contributes to the normal growth of the foetus'.

Comments on the draft guidelines were sought by 31 March 1997, but the announcement of a UK general election saw the FAC review put on hold until after the election result on 2 May 1997. The election result heralded what is widely regarded as a fundamental historic change in UK food policy and the governance of food supply in relation to food safety in the UK. One of the Labour party's election promises was to set up an independent body to protect consumer interests on food safety. The Food Standards Agency will, after much controversy, include nutrition in its remit. The Agency came into being on 1 April 2000, but early indications are that the issue of health claims on foodstuffs will not be taken up as part of its work, much to the surprise of many commentators.

Wide-ranging consumer concerns about functional food

The development of the Joint Health Claims Initiative has to be seen within the context of an unfavourable or even hostile consumer advocacy position on functional food. However, to address consumer issues pertinent to the emerging market for functional food, the National Food Alliance set up a working party at the end of 1995 to develop a reasoned consumer policy response.[2] In a background paper, produced for the working party, 16 key consumer questions that the working party believed needed to be addressed were identified (Winkler, 1998). These were seen as:

- Effectiveness: do functional foods do what they claim to do?
- Volume: must we eat unusual amounts to obtain benefits?
- Superiority: are they better than ordinary products?
- Performance: do they improve mental or physical performance?
- Relevance: is there a need for these products?
- Risks: is there any danger of overdosing on ingredients?
- Substantiation: how should we answer these questions and how much evidence should companies provide?
- Legitimization: do 'good' ingredients just disguise bad food?
- Confusion: can consumers understand and assess claims?
- Price: do these foods cost more and are they value-for-money?
- Claims: how do we prevent exaggerated and misleading claims?
- Labelling: what information should be given on these foods?
- Trade: will varying national regulations limit availability?
- Costs and benefits: do they bring a net gain to society?
- Choice: do functional foods increase real consumer choice?
- Nutrition policy: will emphasis on a single functional food undermine reform of the total national diet?

The Food Commission Attacks Producers Making Health Claims

Consumer concerns about the lack of control of functional food and beverage claims in the UK were highlighted in a damning report published in June 1996 from the Food Commission, the UK's most vocal food consumer advocacy group and a harsh critic of the food industry. A survey of supermarket shelves carried out by the Food Commission found around 800 products making some type of functional claim or implying that they possess a health or nutritional benefit to the consumer. An unexpected conclusion from the Food Commission research is that the functional foods market in the UK seems to be alive and thriving!

The survey included recently-promoted functional products as well as foods which had for some time been promoting themselves as having the benefit of added nutrients or being a rich source of certain nutrients. In some cases the Food Commission survey found examples of products with claims which, under current UK guidelines, would seem to constitute medicinal claims.

Details of the categories of products and numbers making a health claim in the Food Commission survey are reproduced in Table 5.1. The types of dietary benefit or claim made on products are detailed in Table 5.2. In the survey all products which made a claim about the following were noted: vitamin and mineral content (intrinsic or added); bacterial cultures; fibre (intrinsic or added); fish oils or omega 3 (intrinsic or added); fat composition (saturates, polyunsaturates, monounsaturates, hydrogenated fats, *trans* fats and cholesterol); herbal extracts; and suitability for use during exercise.

The survey found that despite the large number of foods promoting their functional advantages, only 12 products referred on their pack to scientific evidence that could underpin the benefits being claimed or implied. In seven cases no scientific reference was given; in three cases there was a reference to a government committee; and in two, to the company's own sponsored research. Only in these two cases did the manufacturer refer to evidence showing the benefits of the product itself (evidence of the benefits of the product sold) rather than the ingredient in general (for example, dietary fibre or fish oil in general).

The report lists four reasons why the fact that a food product claims, or implies a claim, to be of specific health or nutritional benefit should concern consumer and public health organizations:

- The claim may be misleading in that the supposed benefit may not easily be obtained from the product in practical use: the Food

Table 5.1 *Categories of Products and Number Making a Health Claim*

Product	*Making a Claim*
Oats, muesli, cereals (sweet and regular)	149
Baby foods	66
Meat and poultry	56
Fatty spreads	49
Snacks, crisps, biscuits	41
Bread and cake products	37
Oils	35
Fruit juices	34
Frozen vegetables	32
Tinned fish	32
Yogurt and yogurt drinks	31
Dried fruit and pulses	30
Milk: fresh, UHT, evaporated, dried	22
Soft drinks: squashes	22
Soft drinks: carbonates	21
Baby drinks	20
Tinned vegetables	18
Soy foods and drinks	18
Frozen and marinated fish	16
Fruit juice drinks	14
Hot drinks	13
Sweet and savoury spreads	10
Tinned pasta	10
Flour	8
Cheese	4
Mayonnaise	4
Powder potato	4
Tinned soup	3

Source: Food Commission market research, 1996

Commission survey found that most products did not appear to have been tested to ensure that they actually imparted any health benefits to the consumer.

- The supposed health benefit may lead to confusing messages about the nature of a healthy diet. The problem, the report says, of 'junk with added nutrients' is that the nutritional messages about the value of foods become confused and attempts to improve the diet are substantially undermined.
- The foods making the claim may be embedding their supposed beneficial element in a product of poor nutritional quality – a

Table 5.2 ***Types of Dietary Benefit or Claim Made on Products***

Claim	*Number of products*
Contains fish oil/high in omega 3	18
Emphasizes other fats or oils	93
Low in cholesterol	22
Can lower cholesterol	6
Contains bacterial culture which improves digestion	10
Contains bacterial culture which has another physiological benefit (metabolism, dietary balance, immune system, lower cholesterol)	7
High or added source of fibre	232
Vitamin, mineral, vitamin and mineral claim	30
High or rich in protein	92
Other claims and implied claims – isotonic; restorative	36

Source: Bradbury et al, 1996

product high in fat and saturated fat, sugar or salt may be promoted as beneficial to health when in fact such a product 'should be restricted to only a small part of the diet if eaten at all'.

- There are potential safety concerns when nutrients that are hazardous to sections of the population, if consumed at high levels, are being added and/or promoted as beneficial without concurrent warnings. An example, the report says, is found with vitamin A, which is added to some products and promoted in several others, encouraging consumption at levels that, if consumed by a pregnant woman, may increase the risk of birth defects.

The report concluded:

> *'To ensure the best consumer protection, only those health claims which assist the public in choosing a healthy diet – such as the diets promoted by the Health Education Authority, reflecting the consensus view of healthy eating – can be considered to serve a useful purpose. Any other health claims – such as cholesterol-lowering and immune system boosting claims made for bacterial cultures – should be considered medicinal claims and should require the same quality of evidence as a medicinal product must provide. The fortification of food products should be the subject of Department of Health expert review and any claims and promotions on the basis of fortification should be strictly limited.'*

UK Consumers Groups Verdict

The debate on control over food health claims in the UK was given fresh impetus in April 1997 by a report from the National Consumer Council (NCC)[3] stating that health claims confuse consumers more than they help them; they make little sense to shoppers and so don't help them to make healthy food choices and should be strictly controlled. The NCC concluded that an all-out ban would be better.

The report, *Messages on Food*, results from research commissioned by MAFF as part of its review of food claims and describes research into consumer attitudes to health claims on food packaging. The report's findings are based on discussions with groups of consumers from four English towns. The NCC says that the findings highlight the need for a complete rethink on food pack claims and their ability to encourage healthier eating. NCC chairman David Hatch said:

> *'These claims are clearly a waste of space. Many of us are keen to eat a healthier diet, but our report demonstrates just how useless on-pack information is in helping to achieve that. The government and the food industry should look for more effective ways of getting the healthy-eating messages across to consumers.* Messages on Food *also shows that the technical distinction between claims which are regulated, and health claims which are not, is completely lost on most people. "Low fat", for instance, is regulated, but "helps maintain a healthy heart" is not. We now know that this distinction is irrelevant to shoppers. Instead they judge claims by whether they make sense – and few do. Far from helping people choose a healthier diet, our research shows that food claims may actually get in the way of consumers' understanding of what they eat.'*

Some consumers believed that all statements on food packaging are strictly monitored and controlled. Most consumers considered that, if not, they should be.

The research revealed that consumers do not classify food claims in the way that official bodies tend to. According to the authors, consumers place food claims into the following categories:

- Factual – such as 'added vitamins';
- Explanatory – such as 'good for healthy skin';
- Impenetrable – such as nutrition tables;
- Meaningless – chemical or Latin names or terms;
- Spurious – containing words like 'may';
- Unappealing – such as highly chemical or artificial-sounding;
- Esoteric – with long complex statements.

Most health claims – those implying a benefit to health – fell into one or more of the last five categories.

The research contains many pointers for manufacturers thinking about how to label foods. Consumers are influenced, for example, by short one-word or one-phrase claims on the front of food packs. They are also suspicious of longer, more complex claims and find them confusing. In particular, consumers are baffled by the technical jargon and messages about new ingredients. These include omega-3, hydrogenated fats, *trans*-fatty acids and *Lactobacillus bulgaricus*.

Consumers surveyed recognized that their attitudes to simple explanatory messages changed if essentially the same claim contained more information and became longer and more complex. A scepticism about words like 'may' or phrases like 'it is suggested that' was identified in the research. Although consumers recognize that manufacturers have to cover themselves, they found words like 'may', 'believe' or 'suggest' to be unconvincing.

Participants in the research were not naive about food manufacturers and recognized that they should view claims with cynicism. They were, however, confused by many of the more complex health claims. The NCC express concern about the 'fine line between confusing consumers and misleading them' and are worried that claims are getting in the way of official messages about healthy eating. It was also clear that consumers need information about products to be reinforced by other trusted sources and that claims on labels will not change consumers' behaviour unless the message has been reinforced elsewhere.

NCC found that there is a considerable time lag between appearance of information in the public domain and the information consumers come to know and accept. But they say that it would be a mistake to underestimate the slow-drip effect of claims on people's perceptions. NCC calls on FAC to consider whether MAFF should ban health claims on food. If health claims are to be allowed they should be regulated at the UK and EU level. In the interim, they call for voluntary guidelines from MAFF, supported by an enforcement body. They then call upon MAFF to reconvene the discussions on health claims and come up with a set of guidelines.

The negative assessment of functional foods has also been echoed by the UK's Consumers' Association (CA), the publishers of the widely used *Which?* reports. In an article on functional foods published in October 1996, for example, the CA call for better rules to guide manufacturers and to protect and inform consumers. Their conclusion, based on a review of a number of products then currently on the market was stark:

> *'Some of these foods are more worthwhile than others – but our overall verdict is, don't bother with them. Rather than looking for a quick fix from a single product, eating a varied, balanced diet is far more likely to protect your health and prevent disease.'*

The CA also says that food labelling laws need to be updated to cope with all the novel foods appearing on the market.

UK Consumers' Association calls for tighter controls

Consumer concern about functional foods continues. For example, the CA, in a policy report published in June 2000 (Consumers' Association, 2000), called for tighter controls on ingredients, nutrition labelling, and health claims for functional foods. In the report the CA presented their own consumer research on functional foods showing that 'consumers are sceptical about these products and the claims they make'.

Sheilia McKechnie, Director of the CA, said at the launch of the report:

> *'The good news is that most consumers want to eat a healthy diet. The bad news is that aggressive marketing of foods with added health benefits, misleading information on labels, and no mandatory system of approval for health claims means that we can't make sensible choices on food.'*

The CA's report calls for all health claims on functional foods to be proven before the products are marketed. In particular the CA would like to see the UK's Food Standards Agency take a more proactive role, for example by overseeing the Joint Health Claims Initiative, a voluntary process at present. Ultimately, the CA wants action at the European level with a mandatory prior-approval system.

Functional Food Advertisements Fall Foul of Advertising Standards Authority

Most high profile functional food and beverage launches in the UK have been the subject of complaints to the Advertising Standards Authority (ASA), the UK body that oversees print advertising. For example, over a period of three years 21 straight complaints against functional products have been upheld by the ASA, because the claims made for the products lacked adequate substantiating evidence (Winkler, 1998). Table 5.3 lists a number of examples of these.

This series of complaints prompted ASA to issue a strongly-worded warning to food manufacturers, saying 'the strength of the health claims made for some functional foods have begun to raise complaints as well as eyebrows'. The authority emphasized that it would be watching to ensure advertisers took care not to exploit the public's lack of nutritional expertise, and that it expects all claims to be backed by appropriate scientific evidence.

Health Claims Code of Practice

That the UK has voluntarily produced a Code of Practice on health claims on foods in a unique collaboration between the consumer movement, the food industry and food regulators says a lot about the way functional foods have captured the food policy agenda. By way of contrast, for example, there has been no such initiative on animal welfare following the major UK scandals of salmonella in chickens and BSE in cattle!

The JHCI, which produced the Code, was set up in June 1997 and established by the Food and Drink Federation (the food industry's trade body), the National Food Alliance (now called Sustain) and the Local Authority Coordinating Body on Trading Standards (LACOTS). It aimed to keep up the momentum on discussion over health claims that resulted from MAFF's functional foods initiative and the stalling of the FAC review on its draft guidelines on health claims due to the general election.

It was a response to the growing recognition of the role of diet in maintaining good health and an anticipated growth in the market for functional food. The JHCI aimed to also address the need to prevent the use of false, exaggerated, misleading and prohibited health claims which, the Code says, has become a priority issue for the food industry, law enforcement officers and consumers, not only to protect the consumer but also to promote fair trade. In addition, the Code aims

Table 5.3 *Examples of ASA Adjudications on Functional Products*

Case	*Claim*	*Complaints upheld*
Gaio yogurt	Can help lower the level of harmful cholesterol	Cholesterol-lowering effect not sufficiently supported Exaggerated presentation of effects No difference with other yogurts
Mono spread	Monosaturates healthier for the heart than polyunsaturates	Scientific data not conclusive
Ribena juice and fibre drinks	Help lower your cholesterol level...which can increase the risk of a heart attack	Claims that the product could lower plaque build-up on artery walls not supported Exaggeration of health benefits
New Isostar sports drinks	Unique formulation... delivers maximum carbohydrate loading with the speed of isotonic absorption	No proof that the drink is absorbed as quickly as other isotonic drinks
All-Bran breakfast cereal	High in fibre...to remove the toxins...provides the environment in which 'good' bacteria thrive	No evidence to support this
Pact spread	Heart-shaped packaging suggesting 'good for the heart' ...essential omega-3 intake recommended	No proof of omega-3 being good for the heart Omega-3 not essential
Red Bull drink[4]	Suggestion of vitalization/ invigoration/sexual stimulation	Claims not proven or exaggerated

Source: ASA (various years)

to set a consistent approach to health claims not only in the UK, but Europe and internationally.

The central pillar for the Code of Practice to work is the proposal to set up a Code Administration Body (CAB). The draft document from the JHCI sets out alternatives on how best this body could be established.

The Code will be a voluntary process and will not replace or compete with the current systems of regulation and self-regulation and is not concerned with claims about what is contained in a food, only the health benefits that may result. However, the Code will apply to any claim made in advertising, promotion or labelling that a food provides a specific health benefit.

The CAB would be advised by experts to assist companies who want to make health claims and support existing regulatory and self-regulatory bodies. The Code details a set of principles which should be followed when judging whether a health claim is misleading, false or exaggerated. The principles in the Code are applied to the health claim as it appears to the consumer as well as its literal or legal meaning. For example, marketing imagery, careful use of words and literature given out with the product are all taken into account.

The Code makes a distinction between new claims and generic claims. The key part of the Code for new health claims is that these must be proven by the company responsible using evidence in humans, although other evidence, such as animal studies, biochemical and cellular studies and published literature can also be used in support. Importantly, the Code will accept evidence of the effects of foods on established biomarkers – for example, if a food reduces cholesterol, this can be also taken to reduce the risk of heart disease.

For new claims, evidence must show a significant and sustainable benefit as part of the normal diet from regularly consuming a reasonable amount of the food. If possible there should be an explanation of how the product works.

In contrast to new claims, the Code refers to generic claims such as those already well established for vitamins or fibre. These can be used on any product with the right nutrient content without further evidence. It is proposed that the industry develop a list of generic claims for review and agreement under the Code. Until this is complete the Code will only be applied to new claims.

The Code is seen as dynamic and will be kept under review for effectiveness as well as scientific and legal developments in the UK and elsewhere – in particular, the role of the new Food Standards Agency in the UK, developments in international trade and potential European laws on claims.

The draft Code contains three annexes. The first sets out detailed guidance on borderline claims. Advice is also given on how to communicate health claims without breaking the law and on the procedure to follow where an explicit reference to a disease may be needed in order to effectively communicate a health claim to the consumer. The second annex provides advice on setting up trials in support of health claims; and the third sets out options for the Expert Authority and the basic requirements to ensure its credibility.

The UK's Joint Health Claims Initiative (JHCI) was on schedule to be up and running by September 2000. The final text of the Code was given the green light on 14 June and candidates for the post of Executive Secretary for the JCHI interviewed in July. At the same time it is hoped rules for developing 'generic health claims', that is, claims similar to those in the US approved by the Food and Drug Administration, would also be agreed for use in the UK as part of the JHCI process. By June 2000, more than £60,000 ($90,000) had been raised from the UK food industry to help finance the JHCI. The goal is to make the JHCI self-financing through companies paying fees for trying to get 'health claim' approval. It is expected that the health claim application procedure, once a product dossier is in place, would take no longer than 90 days (it is also important to note the JHCI initiative applies equally to dietary supplements as foods and beverages).

The Difficulties of Making 'Health Claims' for Foods and Drinks

Both Nestlé, the world's largest food company, and SmithKline Beecham (SB), which has joined forces with Glaxo to become one of the world's leading pharmaceutical companies, have been found wanting in the UK over claims made on food and drink products. The decisions in the UK that went against both Nestlé and SmithKline Beecham (SB) show what a minefield the question of 'health claims' is when even these responsible and reputable companies can get into difficulties.

In the case of Nestlé, they were found guilty in a UK court of law in May 2000 for making 'medicinal' claims relating to advertising and labelling as part of a Health Heart Campaign run on packs of their flagship breakfast cereal brand Shredded Wheat. While SB were judged in July 2000 by the UK's Advertising Standards Association (ASA), the official watchdog on print advertising, of making a 'misleading' advertising claim for their important beverage brand Ribena ToothKind.

Both companies deny strenuously that they did anything wrong, and in the case of SB, the company is seeking a judicial review to squash the ruling by the ASA. But the decision by the ASA to uphold a complaint against SB sparked off a huge 'health claims' row in the UK.

Two-year investigation

SmithKline Beecham, who launched their innovative children's juice drink Ribena ToothKind in 1998, were told by the UK's Advertising

Standards Authority that part of their packaging claim – that Ribena ToothKind 'does not encourage tooth decay' – is 'misleading' and the ASA asked SmithKline Beecham to delete it from their advertising.

The ruling by the ASA follows a two-year investigation into a complaint made to the ASA about Ribena ToothKind's advertising claim, suggesting that the product prevents tooth decay. The ASA concluded from expert advice that consuming Ribena ToothKind instead of other fruit juices or carbonated drinks carried an oral health advantage but delivered no intrinsic benefit to a child.

The ASA ruling is an embarrassment to SmithKline Beecham and to the British Dental Association (BDA), the professional body representing UK dentists, which lent its name to the marketing and promotion of ToothKind after giving its accreditation to the product. Ribena ToothKind was the first, and remains the only, drink accredited by the BDA. The BDA accreditation and the BDA logo feature on product packaging and the accreditation has been used in advertisements. The claim that ToothKind 'does not encourage tooth decay' also appears on-pack.

Since its launch two years ago Ribena ToothKind has grown into a brand with annual sales of £30 million ($45 million). It is a variant of SB's Ribena brand, which has around £169 million ($253.5 million) annual sales and is the fourth-biggest soft drink brand in the UK. Based on blackcurrant juice (although other flavours are available) Ribena is a decades-old brand.

'War of words' breaks out

As soon as the ASA's adjudication was made public on 12 July 2000, a war of words broke out between the BDA and SB on one side, and Action and Information on Sugars (AIS) on the other. AIS is a voluntary network of health professionals concerned with sugars and public health, and one of the complainants about Ribena ToothKind to the ASA.

AIS immediately called for the drink to be either renamed or withdrawn, saying that 'the very name of the product, as well as the claim misleads parents into thinking Ribena ToothKind will not harm children's teeth'.

SmithKline Beecham for its part issued a statement saying it 'fundamentally disagrees' with the ASA's decision to uphold a complaint against one of its advertising claims and announced it was seeking a judicial review in order to quash the ASA's decision. At the same time, SB demanded that AIS 'retracts or corrects an inaccurate and misleading press release issued in its [AIS's] name'.

Surprisingly, the BDA struck out both at the ASA and AIS, even though the ASA's decision referred to advertising claims made by SB and not to the accreditation by the BDA. In a press statement, the BDA said that it 'challenges' the ASA's ruling against Ribena ToothKind and called into question the ability of the ASA to make judgements on scientific issues of this sort.

In a further, strongly worded statement, the BDA went on to 'denounce' the news release issued by AIS for what it called 'gross inaccuracies' and said it was seeking publication of 'an agreed apology'.

ASA review process criticized

Both SB and the BDA called into question the ability of the ASA, the UK's official and only body which handles complaints about print advertising, to make judgements on science.

The BDA further said that it believed that the fact that ASA had taken over two years to come to a conclusion about the dental health claims 'indicates the difficulty they have had dealing with scientific issues of this sort'.

Two years is indeed an unusually long period for ASA reviews. SB's evidence for its claim was scrutinized by two separate dental experts, both of whom found against the claim. The ASA decision was due to be published back in February of this year, but publication was delayed while, in response to an appeal from SB, the case went to an independent reviewer.

The BDA attacked the ASA and its process in a press statement, saying that the ASA had taken two years over its decision because it was not a scientific body and had had difficulty with the science, adding: 'In its decision, the ASA says that the claim that "Ribena ToothKind does not encourage tooth decay" is misleading. The BDA disagrees and our view is backed up by 1200 pages of peer-reviewed research data.'

The BDA went on to criticize the ASA's choice of dental experts to review the case, saying: 'We were very surprised that they did not take advice from the UK's leading experts on diet and dental disease, simply because these experts also advised the BDA.'

ASA says investigation was thorough

An ASA spokesman in an interview with *New Nutrition Business* (July 2000) said that one of the reasons that the case had taken so long to investigate was that the ASA had taken great care to choose advisers who were independent of the BDA, SB and the complainants who had brought the case. He said: 'We needed to produce a watertight adjudication. We aren't a scientific body but we do have a 38-year track record of adjudicating on advertisements and we have vast experience of using experts to help us with difficult cases – such as with GM foods. It's very rare that we use two different experts – that's a sign of how thorough we have been. We gave our ruling on the advertisement – not on the drink or on the BDA's accreditation – we've no bones with the BDA.'

He continued: 'SmithKline Beecham had two years to present us with the most convincing scientific evidence they could get hold of. We reviewed all the evidence which SmithKline Beecham presented and we're very happy to defend our adjudication. The problem is that the nature of the claim is absolute but the evidence presented is more along the lines of a comparative claim.'

Nestlé found guilty of making 'medicinal claims'

In the case of Nestlé, the world's largest food company was been found guilty in a British court of law for making 'medicinal claims' in a healthy heart promotion for one of its flagship breakfast cereal brands.

In what is now being seen as an important 'test case' on health claims in the UK, the company was found guilty on 23 May 2000 of three charges of contravening the UK's Food Labelling Regulations 1996 and fined a total of £7500 with costs of £13,601.

The charges, brought by Shropshire County Council Trading Standards, related to the advertising and labelling of a Healthy Heart Campaign run on packets of Nestlé Shredded Wheat and Shredded Wheat Bitesize in 1999 as a promotion to raise £250,000 for the British Heart Foundation, the UK's fifth-largest charity.

Nestlé UK, which trades as Cereal Partners in European cereal markets, a 50–50 joint venture with US cereals giant General Mills, pleaded not guilty to all charges.

Significantly, this was the first case in which the UK's Joint Health Claims Initiative Code of Practice was used by the prosecution when considering whether a health claim amounted to a 'medicinal' (prohibited) claim.

The 'test' of the ordinary shopper

In another unusual departure, reflecting the public interest in the case, the Stipendiary Magistrate hearing the case at Shrewsbury Magistrates' Court, Philip Browning gave reasons for his guilty verdict.

In his statement on the verdict, Mr Browning said: 'It was clearly a major marketing campaign linked to the British Heart Foundation. The Food Labelling Regulations make it an offence to sell or advertise any food in respect of which a claim is made that has the property of preventing, treating or curing a human disease or any reference to such a claim.

'Applying what I believe would be the test of the ordinary shopper...in my judgement it is clear beyond doubt that the statements about Shredded Wheat to each of the "campaign steps" invite an irresistible inference that eating Shredded Wheat will reduce the risk of coronary heart disease...'

During the court case it was revealed that between April and August 1999 around 10.5 million packs of Shredded Wheat and Shredded Wheat Bitesize were sold that referred to the 'Healthy Heart Campaign'.

In a statement after the verdict, Cereal Partners said: 'We maintain that last year's promotion was not designed to mislead consumers in any way. The campaign aimed to raise awareness of the risks of coronary heart disease by encouraging people to follow simple lifestyle and dietary advice to maintain a healthy heart and to raise funds for the British Heart Foundation.'

The statement is at pains to point out that the judgement refers to last year's Shredded Wheat Healthy Heart Campaign and 'in no way affects the current Shredded Wheat advertising or packaging'.

In an interview with *New Nutrition Business* (June 2000), David Walker, Chief Trading Standards Officer for Shropshire County Council, said that the court decision is being seen as something of a 'test case' that is already having implications for food manufacturers and their thinking of promoting products in relation to health benefits.

In particular, Mr Walker said: 'I have had a number of communications from people in industry and it would appear as the result of this case a number of companies considering adopting a similar marketing route to Nestlé on the basis of this decision are having to re-think. In this respect this is a very important case.'

Mr Walker also pointed out implications of the case for companies using third party support for marketing. He said: 'Food companies have to exercise great care when they link up with a charitable or other organisation when they are associated with anything to do with disease. It is one matter for a charity to make statements in their literature about disease, but this does not mean it can appear on food labels.'

He also stressed the importance of the Joint Health Claims Initiative Code of Practice which was used by the prosecution to challenge the criteria by which the claims were being made by Nestlé. He said: 'Companies will have to pay full regard to the Code of Practice. Food companies failing to follow the Code could find themselves open to a similar challenge.'

EU Policy Developments on Health Claims

The UK initiative, important as it is, may in fact become redundant before it is really up and running. Many different interests are arguing and lobbying hard for health claims to be allowed as part of EU food regulation. For example, the European food industry, through the European Confederation of Food and Drink Industries (CIAA) has produced a position paper for EU-wide proposals to allow health claims.

The role of the CIAA is to represent the European food industry in discussions on regulation in the EU that affects its members as well as in international organizations. As part of their work, the CIAA produce policy or position papers and have published a number of position papers on health claims.

A revised *Code of Practice on the Use of Health Claims* was published late in July 1999 – it includes the following:

> *'Scope, definitions of enhanced function claims and reduction of disease risk claim; general principles for making a health claim, guidelines for the substantiation and assessment of health claims, guidelines for communication.'*

The CIAA points out that the research in diet-health connections is proceeding and that consumers' focus on the contribution of diet to health is remarkable. It states that health claims on food and the dietary guidance of the authorities and health boards in general can make consumers change to a healthier lifestyle. In its view, the labelling of food – including advertising – is one of the most effective methods of communication with the consumer. The CIAA believes that the investment of resources in the development of innovative products with health benefits will be undertaken and encouraged only if manufacturers are able to communicate these benefits to consumers throughout Europe.

CIAA's basis for permitting health claims is that the products are safe; that the statements are scientifically proved; and that the consumers are not misled.

CIAA recommends that health claims be divided into:

1 Nutrient function claims – Claims that promote the role of a nutrient in the normal physiological function of the body. They are based on well-established and generally accepted scientific knowledge. Ingredients and non-nutritional substances are excluded;
2 Claims related to healthy-eating patterns – These claims refer to official recommendations for healthy-eating patterns, dietary guidelines or similar publications. Several national authorities have published recommendations on healthy-eating patterns and others may do so in the future;
3 Health claims – Claims related to health effects. These claims refer to a specific health-related effect of a food or any of its constituents on the body, on a physiological function or on a biological parameter. Beneficial health effects of ingredients and non-nutritive substances are included. The latest CIAA paper has adopted the terminology corresponding to the latest Codex proposal, that is enhanced function claims. Examples given are 'Calcium improves bone density' 'Product X reduces cholesterolaemia';
4 Health claims – Claims related to the reduction of a disease risk. These claims refer to the fact that the consumption of a food may help reduce the risk of a disease, because efficacy of the food itself, or of the nutrients or other substances contained in the food is known or can be proven. In this case, the disease or disorder is named and the risk reduction is explicitly stated. Examples given are 'Adequate calcium intake may help reduce the risk of osteoporosis in later life' and 'Regular consumption of product X can reduce the risk of coronary heart disease'.

The European Health Product Manufacturers' Association (EHPM), which is an association of federations dealing with health products such as food supplements, herbals and health foods, is of the opinion that there is a need to clarify Directive 79/112 as it relates to their members' products. The European Food Law Association (EFLA), which is an international scientific association mainly focusing on studying food law and contributing to its international harmonization, is finalizing a position paper on health claims. EFLA is of the opinion that disease risk reduction claims should be allowed by amending Directive 79/112 and Directive 65/65, rather than introducing completely new legislation for this purpose.

In contrast, the main European Consumer Association – Bureau European des Union de Consommateurs (BEUC) – is basically against health claims. In its view a single food product is neither healthy nor unhealthy; only the diet can be healthy or unhealthy. However, as health claims are now being used in the EU, primarily generic claims on the connection between diet and the development of diseases are recommended, such as the generic claims approved by the US Food and Drug Administration (see Chapter 7).

BEUC's perception is expressed as follows:

> *'The health benefits of a balanced diet cannot be attained through the consumption of food supplements, fortified foodstuffs, functional foods or phytoactive substances alone. The role of these products in the improvement of the diet can only be supplementary. Within the EU, there is an urgent need to adopt dealing with nutritional and health claims. This legislation could be based on the existing legislation in place in the USA.'*

BEUC has been working on a new position paper on health claims. Experts are predicting that there will be provisions within the EU to communicate the health benefits of functional food to consumers in less than two years. This will be achieved through adapting the Food Labelling Directive so it becomes lawful to mention the disease risk reduction properties of functional food. It is not yet clear what the detail of such change might be or if the timetable of change is realistic. Further, at the international level, Codex Alimentarius is well into the process of drafting guidelines on health claims that again could supercede any national initiatives. While the current policy environment is uncertain, the drive for health claim regulation has been gaining momentum in recent years and is in the process (by regulatory standards) of fairly rapid review and change. Expert commentators see these developments as resulting in regulatory and policy provisions being made to allow disease risk reduction claims for foodstuffs.

The situation in Europe is thus one of change – in terms of legislation as well as administrative practice for health claims. There is general agreement that scientific evidence exists to back claims for reduced risk of many diet-related diseases, but there is a wide range of opinions as to whether this knowledge should be permitted to reach customers through advertisements and health claims for food. Critics refer to the fear of misleading consumers, whereas advocates refer to the contribution towards reducing the impact of many diseases of modern civilization, and to the development of innovative, healthier foods. In the next part we examine what is going on in practice, using Japan, the US and Finland as specific examples.

Part 3

The Unfolding of the Functional Foods Revolution Across the World

Chapter 6

The Japanese Invented Functional Food

We would like to introduce you to Mrs Nagayama. By 1999, she had been visiting offices in the Chiyoda district of Tokyo for 21 years selling functional foods face-to-face to office workers. Mrs Nagayama, a mother of two sons in their 20s, is one of Japan's star Yakult ladies, part of a unique door-to-door distribution network first started by Yakult Honsha in 1963, that now numbers more than 53,000.

Yakult Honsha is probably unique in the world of food and health. Decades before healthy eating or functional foods were even thought of, Yakult's founding father, Dr Minora Shirota, was putting his vision of good health through diet into practice. The company he founded, first incorporated in Japan in 1955, is built entirely upon the health benefits of live probiotic lactic acid bacteria which have been shown to help keep intestines in a 'good, healthy condition'.[1] In the case of Yakult Honsha the lactic acid bacteria is called *Lactobacillus casei Shirota*, named after Dr Shirota, and sold as a fermented milk drink in tiny 65 ml bottles. Dr Shirota, a medical doctor by training, discovered the properties of beneficial gut bacteria in the 1930s while working among the poor and malnourished in Japan.

As well as his medical work, he developed strong convictions about health and diet, a philosophy that still guides the company (Heasman, 1999b). For example, Dr Shirota's spirit is kept alive in Yakult through people like Mr Hirokatsu Hirano, a main board director. When we met Mr Hirano in 1999 at Yakult's head office in the Ginza district of Tokyo, he explained to us how as a young man he carried Dr Shirota's medical bag. As they travelled together Dr Shirota would tell him how people should have the power to look after their own health. Dr Shirota's philosophy is straightforward – prevent disease rather than treat disease; a healthy intestine leads to a long life; and deliver health benefits to as many people as possible at an affordable price. As Mr Hirano told us, it is Dr Shirota's philosophy on the meaning of human health that has made and continues to make Yakult a unique company.

The Amazing Yakult Lady

But back to Mrs Nagayama. When we joined Mrs Nagayama on one of her morning rounds we visited an electrical engineering company called SunTec. We marvelled as she went from floor to floor, desk to desk, knowing exactly what each person wanted from her ice-chilled delivery bag. Even when the desks were empty she left her customers' normal product choices for them to find on their return. She also took time out to collect money for the products, often helping herself from people's purses, a reflection of the type of customer relationship she had, one built on loyalty and trust. In this company we also noticed a strange hierarchy of consumption. The company president had a fondness for Yakult's *bifidus* with calcium product and it was fascinating to watch as all the senior managers took the same product, but none was taken by the junior employees. Pushing her cart loaded with produce around the streets, Mrs Nagayama is constantly greeted by passers-by, and by return she has a smile for everyone. In her hard-working, eight-hour day she can visit up to 60 offices. It is clear to see why Mrs Nagayama won one of Yakult's coverted awards for outstanding sales in Japan.

She is based at Yakult's Kojamichi Centre in north-west Toyko, one of 139 Yakult marketing centres in Japan. At the time we met her she was one of 19 Yakult ladies based there. Most of the ladies were new to the centre, but one Yakult lady had served even longer than Mrs Nagayama, with 22 years' service – two others had worked for 19 and 21 years. The centre served a crowded office district with a working population of over 30,000. When we visited the Kojamichi Centre it was achieving daily sales of 2472 dairy-based products. Mrs Nagayama was the centre's top Yakult lady with sales per day of 363 dairy products (Yakult Honsha measure all their sales throughout the world in terms of bottles per day and then equate this to a country's population; targets of 'bottles per day' are part of the company philosophy).

Throughout the world, products are sold using the 'Yakult ladies' system to spread the message on gut health, except in European countries, where instead Yakult uses extensive face-to-face sampling campaigns through supermarkets. In Europe there is just one product on sale at the start of 2000 (see Chapter 9). But in Japan, Yakult have developed a more extensive range of functional products; these are listed in Table 6.1, some of which may make their way to international markets. In addition to its functional products, the company is now a major producer of soft drinks, which are also sold through the Yakult ladies system. Also extraordinary is that over the years Yakult Honsha has identified more than 1000 Yakult imitators or lookalikes, with 300

Table 6.1 ***Yakult's Japanese Range of Functional Foods***

Product	*Description*	*First introduced*	*Lactic acid bacteria profile*
Yakult 400	Fermented milk drink	1999	40 billion lactobacilli
Yakult	Fermented milk drink	1935	6.5 billion *L. casei Shirota* strain
Yakult 80 Ace	Fermented milk drink	1991	30 billion *L. casei Shirota* strain
Joie	Drinking-type yogurt	1970, 1993 (re-launch)	12.5 billion *L. casei Shirota* strain
Mil-Mil	Fermented milk drink	1978	10 billion *Bifidobacterium breve Yakult* strain; *Bifidobacterium bifidum Yakult* strain; *Lactobacillus acidophilus*
Mil-Mil E	Fermented milk drink	1982	10 billion *Bifidobacterium breve Yakult* strain; *Lactobacillus acidophilus*; *Streptococcus thermophilus*
Bifiel	Fermented milk drink	1989	1 billion *Bifidobacterium breve Yakult* strain; *Streptococcus thermophilus*
Whip Land	Whipped-type yogurt (like mousse when chilled; like ice cream when frozen)	1993	1 billion *Lactobacillus acidophilus*
Sofuhl	Solid-type yogurt	1975	10 billion *L. casei Shirota* strain

Source: Yakult Honsha, 1999

competing products in Japan alone, but it still retains a strong leadership position in many countries.

Another unusual feature about Yakult Honsha, again inspired by Dr Shirota's philosophy, is its internationalization. Yakult is active in the following countries: Republic of China (Taiwan) since 1964; Brazil

1968; Hong Kong 1969; Thailand and Republic of Korea 1971; Philippines 1978; Singapore 1979; Mexico and Guam 1981; Indonesia 1990; Australia 1994; The Netherlands 1994; Belgium and Luxembourg 1995; UK and Germany 1996; and Argentina 1997.

As can be seen, Europe has only recently received the Yakult 'philosophy' and we explore the market implications of this activity more fully in Chapter 9. Based on the science and technology of the health benefits of probiotics, the company claims 26 million bottles of Yakult are drunk in the world every day (14.2 million bottles of Yakult a day internationally in March 1999). Yakult Honsha's total business in the financial year ending March 1999 had sales of Y152,588 million (US$1.3 billion). Of this total, 54 per cent was derived from sales of dairy products, 32 per cent from juices and soft drinks, 4.4 per cent from cosmetics and 3.9 per cent from pharmaceuticals (both of these last two businesses being derived from Yakult's expertise in lactic acid bacteria), and 4.8 per cent from other activities.

Yakult Steps In to Champion the FOSHU System

As Yakult steps into the new century it has taken on a new marketing route putting it at the heart of modern-day functional food developments. In Japan the company is becoming a driving force in Japan's unique FOSHU system. Foods for Specified Health Uses (FOSHU) are defined as those to which a functional ingredient has been added for a specific healthful effect and are designed to maintain or promote good health. Japan has become internationally known for this system for the regulation of functional food, and we describe this in detail below. In the eyes of the Japanese, the difference between FOSHU and functional food is that FOSHU products have health benefits based on approved scientific research data, including clinical trials. This innovative and ground-breaking system could serve as a model in many Western countries for allowing health claims on individual functional foods.

Less well known however, despite being introduced in September 1991, is the FOSHU system's relative lack of recognition in the Japanese marketplace. In 1999 less that 15 per cent of Japanese consumers were aware of FOSHU (JHFNFA, 1999). To help change this situation the Japanese government approached Yakult Honsha suggesting it put its products through the FOSHU system for approval to become a champion for FOSHU. A number of Yakult products were approved in 1998 and in 1999. Yakult FOSHU-approved products made up more than 38 per cent of all FOSHU-approved product sales (total FOSHU sales were Y227 billion in 1999, around US$2.1 billion).

Yakult is now cooperating with the Japanese government to communicate more effectively to the consumer what FOSHU means. The Yakult products joined a number of other dairy products with FOSHU approval. These include:

1 Lactic acid bacteria fermented milks:
Healthy Family, Frozen Yogurt (Fructo Oligo Grico company)
Takanashi Drink Yogurt LGG, (Tananashi Dairy)
Bifidus Plain Yogurt, *B. longan BB536* (Morinaga Dairy)
Meiji Bulgaria Yogurt, *L. bulgaricus 2038*, *St. thermophillus 1131* (Meiji Dairy)
Snow Brand Nature Yogurt, *L. acidophilus*, *B. longum STB 2928* (Snow Brand Dairy)
2 Other fermented milk drinks:
Yogurina, xylo-oligosaccharide (Suntory Co Ltd)
Calpis Ameals, Lactotripeptide (Ajinomoto Calpis Co Ltd)
Echo Life, isomalt-oligosaccharide (Mil Honsha Co Ltd)
Bowel Friendly, oligosaccharide, fructo-oligosaccharide (Kyodo Dairy)

In short, a large part of Japan's functional food industry is based upon a mixture of new and very traditional foodstuffs and ingredients.

The Regulation and Marketing of Functional Food in Japan[2]

Yakult is just one of more than 300 Japanese companies actively involved in FOSHU and functional food developments (these are now two distinct groups of products which are explained below) making the Japanese market for functional food the most sophisticated in the world. There are two unique features about the Japanese market for functional food. First, the range and diversity of the products on offer, and, second, the role government has played in initiating and promoting the environment in which functional food can flourish.

The concept of functional foods was first invented in Japan in 1984 by scientists studying the relationships between nutrition, sensory satisfaction and fortification and modulation of physiological systems – the functional elements of foods (Hosoya, 1998). But although the Japanese invented the concept of functional food they then dropped the term in 1991 in favour of FOSHU so as not to confuse functional foods with pharmaceutical regulation. Uniquely, they then created a special food category for FOSHU (functional foods), something Western policy makers and industry leaders have been reluctant to propose. The separate FOSHU category became one of five categories of foods regulated for special dietary uses according to Japan's nutrition improvement law. The other foods regulated for special dietary uses are foods for the sick, powdered milk for pregnant or lactating women, formulated powdered milk for infants and foods for the elderly.

The First Functional Food

The product credited with being the first Japanese functional food is a dietary fibre soft drink called Fibe Mini from Otsuka Pharmaceutical, first launched in 1988. This is still a 'hit' product in Japan. It uses the soluble dietary fibre polydextrose as its functional ingredient and is marketed for 'gut regulation'. Fibe Mini started Japan's functional foods goldrush and between 1988 and 1989 51 dietary fibre drinks products designed to aid gut regulation were brought to the market. Functional foods and beverages for gut regulation dominate the Japanese market (see below). Since Fibe Mini hundreds of functional products have been launched in Japan and later in this chapter we present an analysis of the trends in those launched between 1988 and 1998.

Functional ingredients used in commercial products for different health reasons range from the mainstream to the frankly bizarre. For example, there are products using many types of oligosaccarides as prebiotics; a variety of sources of dietary fibre to regulate the gut; minerals such as calcium, iron and mineral absorption promoters; essential fatty acids (mainly DHA and EPA to activate the brain); carotene to prevent cancer and other diseases of ageing; proteins and peptides to control hypertension or cholesterol; lactic acid bacteria cultures to improve gut health and prevent growth of intestinal pathogens; non-cariogenic sweeteners; polyphenols and other substances which prevent dental caries; collagen for skin care; phyto-chemicals to relieve eyestrain; mushroom extracts to prevent 'bad breath'; chitosan to reduce cholesterol; green tea catechins to prevent cancer and health teas for a variety of reasons including anti-allergy.

In comparison, vitamin-fortified foods (with the exception of carotene – the vitamin A precursor) are not considered to be functional foods in Japan. Many functional foods, however, are also fortified with vitamins, but products only with vitamins are not seen as functional.

There are two routes to market for manufacturers to take. First, is the (voluntary) FOSHU approval system, which allows product-specific health claims to be made if a product is approved – 174 products had gained FOSHU approval by February 2000. The second, and more common, is the non-health claim or non-FOSHU route, where companies simply launch products without making any explicit health claim. More than 2000 products have taken this way to market, thus demonstrating that there is a flourishing market for non-FOSHU functional foods.

Launching a product without a health claim can prove very successful because of the high level of knowledge among Japanese consumers about food and health. Companies use marketing communications and packaging very skilfully to put their health message across to consumers. Consequently, products without health claims (that is, without FOSHU approval) often do extremely well. For example, 'The Calcium' is a calcium-fortified sandwich biscuit from Otsuka Pharmaceutical. There are no specific claims for it, apart from the information that one biscuit contains 600 mg of calcium – there is just the general phrase on the front of the pack which states:

> 'The Calcium, *the delicious cream sandwich to build and keep a healthy body – from children to aged people. Do not forget to exercise and eat* The Calcium – *every day.'*

Otsuka has avoided the need for FOSHU approval by not making any specific health claim. 'The Calcium' has become a hit product and is probably the most successful of all calcium-fortified foods in Japan.

The advantage to companies taking the FOSHU route is that approval allows product-specific health claims to be made – FOSHU are the only types of food that can carry health claims. In addition, product packaging can use an attractive logo which reads 'Ministry of Health approved food for specified health use'. It also allows companies to differentiate their products on health grounds in a crowded market. However, while gaining in popularity, FOSHU products only account for around 10 per cent of Japan's total functional food and beverage market. FOSHU products had a market value of Y227 billion (US$2.1 billion) in 1999, which suggests that the total functional food market in Japan is about US$10–11 billion.

An Informed but Changing Consumer

A number of factors have contributed to the demand for functional foods in Japan. Before the functional food boom there had been an expanding market for health foods in Japan for many years. The

Japanese consumers are health-conscious and well-informed. In addition the market for food and drink in Japan is very competitive and a new food can suddenly become a hit product and just as suddenly disappear. There is great competition for shelf space in the stores, in particular in the convenience stores (small local stores, which stay open long hours). If a product is to become a hit then it must get onto the shelves of convenience stores. The store manager will only agree to take a product if it has hit potential. This vicious circle can only be overcome by launching something truly novel.

Japanese consumers usually shop every day. This is partly because Japanese homes are very small, but also because Japanese people like to eat something different for every meal and they are willing to try new foods. This emphasis on variety is even reflected in Japan's dietary guidelines. For example, in 1985 the Japanese Ministry of Health and Welfare issued *Dietary Guidelines for Health Promotion*, recommending that up to 30 different foods per day should be consumed (Matsumura and Ryley, 1991).

The consumer in Japan is accustomed to a wide choice and to seeing something new on the shelves almost daily. The market moves very quickly. If a new product idea becomes a success, all the top companies launch their 'me-too' versions with amazing rapidity – sometimes within weeks. The food industry, therefore, is always looking for new ideas to attract the attention of very fickle consumers. It was against this background that the concept of a functional food was born. But although functional food product development in Japan is sometimes seen as less relevant to the West, this misses the international influences also shaping Japanese food culture, which one researcher notes as the culinary globalization of Japan (Cwiertka, 1998 and 1999). Western culinary ideas have had a great deal of influence on the Japanese diet, even to the extent that researchers are noting decreasing differences between Japanese and American death rates from nutritionally related diseases (Lands et al, 1990). In many respects Japan's functional foods may be a lot less Asian than some Western companies might assume. For example, while we were in Tokyo, the trendiest eating-out cuisine was Italian!

The Role of Government in Developing the Japanese Market

The Japanese government, like Western public policy interventions, has been active in promoting the role of diet in maintaining good health and in disease prevention. For example, Japan introduced its own healthy-eating dietary guidelines in 1985. The government also became increasingly concerned about the health and well-being of its ageing population – Japan having the highest average life-expectancy in the world. From these concerns the concept of functional food was developed and implemented.

The role of the state in food regulation in Japan is based on a different approach from that in the West, for example compositional standards are often voluntary. But, often overlooked in the West, is the extent of the role of the Japanese government in facilitating the development of the functional food and beverage market. In fact the Japanese government has played a pivotal role, starting in 1984 to 1986 with the setting up of a special study group on functional food in the Ministry of Education, Science and Culture; initiating an approval system for functional food; and, importantly, through the active identification and promotion to consumers of the need to correct dietary deficiencies.

For example, among the concerns of the Ministry of Health and Welfare is the deficiency of dietary fibre in the Japanese national diet. A report from the Ministry in December 1989 showed that the fibre content of the national diet had declined over the previous 35 years from 22.7 to 17.37 g/day in 1985. The recommended daily intake of dietary fibre is 20–25 g/day. The Ministry had also been warning for some time that the national diet was deficient in calcium and this was a danger to children in particular. They published figures showing that the calcium content of the national diet was 552 mg per person per day in 1975; this had fallen to 539 mg/day in 1992, whereas the recommended daily intake is 600 mg/day. Much publicity was given to this concern over a lack of calcium. Consequently, consumers readily accepted the benefits of products fortified with calcium and fibre.

In response to such national dietary deficiencies, the Japanese government has encouraged the development of functional foods and beverages in relation to public health. This had gained momentum in the mid-1980s when, in 1986, the Ministry of Education completed an initial report on functional food. In the same year, the Ministry of Health and Welfare established the Functional Food Forum to investigate ways of improving the population's health. In 1988, the Japanese Association of Health and Natural Food and the Research Board of Technology for Future Foods were established and the Ministry of Health and Welfare set up a functional foods committee. In 1989 the latter committee submitted its interim report and the government then established the Functional Food Liaison Board to serve as sole intermediary with industry. At the same time, individual subcommittees were formed to deal with each potential functional ingredient.

These public policy initiatives leading to the introduction of the FOSHU system are as follows.

1984	The concept of functional foods established by Japanese scientists;
1988	Establishment of Functional Food Panel in the Ministry of Health and Welfare;
1989 (April)	Publication of interim report on functional food regulation;
1990 (March)	'Legislation Panel on Functional Foods' established;
1990 (November)	Drafting of report on the legislation of functional foods;
1991 (July)	FOSHU became law after revision of the Nutrition Improvement Act (enacted September 1992);
1991 (Sept)	FOSHU approval system for health claims on foods established.

The FOSHU Approval System for Functional Food and Beverages

FOSHU approved foods are now a distinct category within the Japanese food supply and it is a system aimed at healthy people. The definition and description of FOSHU products is given as a detailed note in Appendix 1. With this new system the term 'functional' has been dropped following the issuing of the decree by the Ministry of Health in July 1991 under the Nutritional Improvement Law. There are also requirements to print warnings regarding intake of some of these foods. But it is important to note that it is the complete product as consumed that receives FOSHU approval, not isolated nutrients or ingredients, such as oligosaccharides and dietary fibres. This confusion is a common error in analysis of the FOSHU system in the West.

The FOSHU system was slow to take off. Only 13 products were approved in 1993–94; 41 products were approved in 1999 and the total of approvals reached 174 by February 2000.

This in part can be explained by a revision of the FOSHU system at the end of 1997 that simplified the approval process making it much shorter. It is interesting to note the changes to the FOSHU system that came into effect early in 1998. The main changes are:

- Removal of the time limit. Previously a FOSHU approval was for a period of four years only. In order to protect the consumer, the

quality control inspection system is to be strengthened, with quality control checks at the manufacturing plant by local authority food law inspectors;
- Documentation reduced and simplified;
- Series of products. Previously a separate application had to be made for each product in a series. Now, if the only difference between similar products in a series is, for example, the flavour or aroma or the addition of minor ingredients (processing aids), then one application can cover a whole series;
- Shorter application procedure. The procedure has been shortened by removing obligatory testing by the National Health and Nutrition Laboratory from the process. The manufacturer's own analytical testing will suffice;
- Change application after approval. It is now possible to make changes in the scope of a FOSHU approved product after approval has been granted, if these changes are of benefit to the consumer.

Gut health is the still most important category of FOSHU approved products, accounting for 73 per cent of 167 FOSHU approvals up to the end of September 1999. In terms of value, gut health is even more important, representing 82 per cent by value of the Y227 billion FOSHU market in 1999.

Other categories of FOSHU approved products for specified health use include hyperlipemia (19 products); mineral absorption – Ca/Fe (eight); hypertension (four); tooth caries (five) and hyperglycemia (two).

Although it is products that receive FOSHU approval, it is specific ingredients that are delivering the specified health use. An astonishing range of different ingredients is being used for FOSHU products. Table 6.2 presents an analysis of 154 FOSHU approved products of the functional components and the types of foods and beverages where they are being used. What is striking is the range of ingredients and products delivering health benefits to consumers. There are, however, some ingredients more prominent than others. Of note are:

- The 12 products using fructo-oligosaccharides, especially their use in seven table sugar applications;
- The eight products using soy bean oligosaccharides (approved for 'good intestinal environment');
- The seven products that have been approved using soy protein, in particular meat products (different types of sausage);
- The 14 products using lactosucrose (for increasing bifidobacteria living in the gastrointestinal tract (GI);

Table 6.2 *FOSHU: functional components and product types*

Functional component	*Type of food*
Xylo-oligosaccharides	Lactic acid bacteria drink
	Chocolate
	Candy
	Seasoning vinegar
Calcium citrate malate (CCM)	Soft drink (x3)
Fructo-oligosaccharides	Table sugar (x7)
	Tablet candy
	Candy
	Ready-to-eat pudding
	Soft drink
	Lactic acid bacteria drink
Soy bean oligosaccharides	Soft drink (x3)
	Carbonated beverage
	Table sugar (x4)
Isomalto-oligosaccharides	Table sugar
	Carbonated beverage
	Lactic acid bacteria drinks
Soy protein	Soy-bean balls for frying
	Fermented soy protein drink
	Meatball
	Hamburger
	Soft drink
	Wiener sausage
	Frankfurt sausage
Lactosucrose	Soft drink (x6)
	Frozen yogurt
	Table sugar (x4)
	Candy
	Biscuit (x2)
Casein phosphopetide (CPP)	Soft drink (x2)
	Tofu
Lactulose	Soft drink
Galacto-oligosaccharides	Table sugar (x3)
	Seasoning vinegar
Indigestible dextrin	Wiener sausage
	Soft drink (x2)
	Bologna sausage
	Frankfurt sausage
	Nata de coco
	Powdered-type soft drink
	Cookie

Functional component	*Type of food*
Polydextrose	Carbonated beverage
	Soft drink (x2)
Partially hydrolyzed guar gum	Powdered soft drink
Palatinose/green tea polyphenols/ maltitol[a]	Chocolate (x2)
Heme iron	Chewing gum
	Candy (maltitol) (x2)
Lactobacillus GG	Soft drink (x2)
Glycoside from Eucommia leaves	Yogurt drink
	Yogurt
Casein dodecapeptide	Soft drink (x2)
Chitosan	Soft drink
Psyllium husks	Biscuit (x2)
	Powdered soft drink (x7)
	Instant noodles (x6)
	Cornflakes (x2)
Low molecule sodium alginate	Soft drink (x3)
Bifidobacterium longum BB536	Yogurt (x2)
Lactobacillus delbrueckii (subsp. Bulgarius 2038) and Sreptococcus salivarius (subsp. Thermophilus 1131)	Yogurt (x6)
Lacto-tripeptide	Lactic acid bacteria drinks
Lactobacillus acidophilus SBT–2062 and bifidobacterium longum SBT–2928	Yogurt
Dried bonito oligopeptides	Soup powder
	Powdered miso soup
Lactobacillus casei Shirota	Lactic acid bacteria drink
	Fermented milk drink (x4)
	Yogurt (x16)
Bifodobacterium breve Yakult	Yogurt (x3)
Diacylglycerol	Cooking oil (x4)
Diacylglycerol phytosterol (beta-sitostanol)	Cooking oil (x4)
Wheat bran	Cereal (x2)
Protease hydrolysate	Soft drink
'Sardine peptide' (valyl-tyrosine)	Soft drink
Raffinose	Soup

Notes: Based on 154 FOSHU approvals to end of September 1999
a Used in combination to prevent tooth decay
(x2 etc) equals the number of products using this ingredient
Source: JHFNFA, 2000

- The different types of product (eight in total) using indigestible dextrin;
- That one of the most widely used ingredients is psyllium husks, being used in 15 approved products (used as a rich natural fibre to help regulate and maintain a good GI condition);
- That there have been a large number of approvals for products using probiotic lactic acid bacteria, notably the 21 products using *Lactobacillus casei Shirota*.

It is therefore, interesting to note the mix of old and new functional ingredients. Many products, such as those using probiotic lactic acid bacteria, were well established in the Japanese market before the invention of functional foods in the late 1980s. But these are now leading the market having been given new positioning and a new lease of life as FOSHU-approved foods.

The Functional Food Market is More than Just FOSHU

While the FOSHU market has been accelerating in recent years, by value and number of product approvals, this still represents less than 10 per cent of all functional products launched in Japan. Our analysis (Table 6.3) shows that about 1720 functional products were launched between 1988 and 1998.

As can be seen from Table 6.3, functional food product launches have only accelerated in recent years. More than 55 per cent of product launches took place between 1996 and 1998 with around 42 per cent in the past two years. What is also striking is the wide range of different foods and beverages in different food categories that have been launched. In this sense functional foods in Japan are part of everyday, mainstream food supply. While soft drinks are important as a single category (23 per cent of all product launches), other product categories have taken significant shares, in particular milk drinks, lactic acid bacteria drinks and yogurt products.

There are four broad categories of functional ingredients, namely, dietary fibres, oligosaccahrides, calcium, and, to a lesser extent, the essential fatty acids DHA and EPA (Table 6.4). Of these calcium has been the most widely used single ingredient, with 53 per cent of all products using calcium being launched between 1996 and 1998. This is followed by dietary fibres (42 per cent) and oligosaccharides (37 per cent), which are both used for gut health. This usage matches government advice to consumers on dietary deficiencies – that is, lack of fibre and calcium.

Table 6.3 *Functional Foods and Drinks Launched in Japan (1988–98)*

	1988	*1989*	*1990*	*1991*	*1992*	*1993*
Soft drinks	29	28	33	11	16	13
Milk drinks and lactic acid bacteria drinks	2	9	15	3	27	21
Yoghurt and drinking yoghurt	3	14	17	5	5	9
Sweeteners	0	0	7	0	1	2
Breakfast cereals	0	3	4	3	2	0
Confectionery and biscuits	2	9	11	13	5	8
Others	2	11	37	7	22	26
Totals	**38**	**74**	**124**	**42**	**78**	**79**
	1994	*1995*	*1996*	*1997*	*1998*	*Totals*
Soft drinks	30	28	39	46	123	396
Milk drinks and lactic acid bacteria drinks	19	25	28	35	48	232
Yoghurt and drinking yoghurt	13	18	35	41	63	223
Sweeteners	7	5	4	0	9	35
Breakfast cereals	1	1	2	2	2	20
Confectionery and biscuits	29	35	23	50	60	245
Others	58	73	86	128	120	570
Totals	**157**	**185**	**217**	**302**	**425**	**1721**

Note: 'Others' includes meats, snacks, refreshments, sauces, dressings, spreads, desserts, ice cream and baby foods
Source: Japanscan Food Industry Bulletin and *New Nutrition Business*, 1999

Conclusion

While functional food has become a firmly established part of Japanese food supply, there is still a long way to go before they can be

Table 6.4 *Leading Functional Ingredients in Japan (1988–98)*

	1988	*1989*	*1990*	*1991*	*1992*	*1993*
Dietary fibre	20	46	57	12	18	14
Oligosaccarides	9	34	54	12	18	14
Calcium	4	18	41	21	37	37
DHA and EPA	0	0	0	0	1	2
Totals	**33**	**98**	**152**	**45**	**74**	**67**
	1994	*1995*	*1996*	*1997*	*1998*	*Totals*
Dietary fibre	21	21	39	50	63	361
Oligosaccarides	20	26	38	49	31	305
Calcium	69	85	96	115	142	665
DHA and EPA	31	27	11	7	5	84
Totals	**141**	**159**	**184**	**221**	**241**	**1415**

Source: New Nutrition Business, 1999

described as mainstream. But it will be important to follow how Japan's leading functional food companies help grow and develop this market, especially FOSHU, and whether others will take Yakult Honsha's lead and look to overseas markets for expansion.

In the meantime, there are important lessons to be learnt from the Japanese experience, not least the fact that many European and American companies have failed to fully understand the concept of functional foods as practised in Japan. Japan's functional foods, old and new, are often truly novel product innovations. Such products offer incremental growth and are rarely simply 'health benefit' line extensions. Most products are also geared to prevention, that is keeping healthy people healthy, rather than targeting disease states. Clearly the Japanese market and its development is very distinct in that it is geared for domestic circumstances. However, few Western companies to date have managed to emulate the success of functional food in Japan or adapt the concept to Western market and cultural conditions. The main trends in the functional market in Japan can be summarized as:

- More companies are taking the FOSHU route to differentiate their products. This now accounts for more than 10 per cent of the total functional foods and beverages market by value;
- The focus of many products has been on recognized and documented population-wide dietary deficiencies, for example calcium and dietary fibre;

- The focus is on foods and beverages, not dietary supplements, tablets and capsules;
- Two distinct paths are being followed both with successes and failures: the FOSHU path which allows product-specific health claims and the non-FOSHU path which has achieved greater than 90 per cent of product launches;
- There is a sophisticated consumer interest and awareness of lifestyle products, including food and health products. Added to this is the often innovative product design and packaging to 'deliver' the health benefit and capture the imagination of consumers;
- The Japanese government has played a pivotal role in developing, promoting and creating the climate for functional foods to thrive. This has included setting up and supporting a wide-ranging functional food research programme and a self-regulatory product approval process that allows product-specific health claims.

Chapter 7

The US – Breaking the Boundaries between Food, Medicine and Health

If there was a starting point for the functional food and nutraceuticals revolution in the US, then it has to be October 1984. This was when The Kellogg Company used advertisements and explicit package labelling claims on its All-Bran breakfast cereal to inform consumers that a high-fibre and low-fat diet could reduce the risk of developing certain forms of cancer. With its history of nutritional rebellion it seems only natural that it was Kellogg that lit the fuse that has, unintentionally, led to the health claims anarchy confronting the American consumer.

But this starting point is grounded very much in the context of the healthy-eating revolution described in Chapter 3; it also forms part of a long history in the US of food manufacturers attempting to make health claims in relation to their products (Hutt, 1986). However, the action taken by Kellogg to market the health promoting benefits of All-Bran has all the characteristics of the functional foods revolution which it predates by almost a decade.

Importantly, adding to the credibility of this promotion was a third-party reference – the National Cancer Institute (NCI) – which reviewed Kellogg's fibre-cancer claims prior to the launch of its All-Bran cereal promotion. The NCI was used in promotional materials to endorse the Kellogg campaign. At the time, the Kellogg promotion was also illegal, being in direct contradiction to FDA policy that had banned the use of health claims to promote food products. In fact Kellogg had bypassed the FDA – for example it did not first check its statement about the relationship of dietary fibre to cancer with the FDA although it did with the NCI (Hutt, 1986). But the Kellogg campaign, in terms of public policy, was the stimulus for a full review of the use of health claims in food promotion (Hutt, 1986; Ippolito and Mathios 1990a) and resulted in the ban on health claims in the US being suspended while this took place.

Kellogg's stance sparked an explosion of unregulated health and disease-related claims as other manufacturers sought to emulate

Kellogg's lead. For example, the *New York Times* documented that in the first half of 1989, 40 per cent of the new food products introduced bore a health claim (reported in Sims, 1998) and magazines like *Business Week* brought public attention to the ridiculous nature of some health claims with a lead story in 1989 headlined 'Can Cornflakes Cure Cancer?'

But the Kellogg campaign had a significant nutritional as well as market impact. After 1984, per capita consumption of fibre in cereals shot up to double its previous levels, mainly because manufacturers started to increase the fibre content of their cereal products. As Ippolito and Mathios (1990) say:

> *'Cereals introduced between 1985 and 1987 were significantly higher in fibre than the average cereal on the market in 1984...Thus, the evidence on new product developments in the cereal market is consistent with the hypothesis that the ability to use health claims to advertise new products is a significant factor in stimulating the development and introduction of more nutritious products in the market.'*

Using a simplified model of the cereal market, Ippolito and Mathios estimate that the market for high-fibre cereals increased by US$280 million over market projections in 1987, which implied an increase of approximately two million households eating high-fibre cereals due to the advertising. The US functional food and nutraceuticals market was starting to take shape.

The critical lesson of Kellogg's actions relates to the role of information. In September 1984 All-Bran was just an ordinary box of breakfast cereal but by October 1984 All-Bran was being seen as possibly preventing cancer! It was still the same ordinary box of All-Bran; the only difference was that information had been added to the All-Bran image, in this instance a statement about the relationship of dietary fibre to cancer. For the new millennium, the role of information is even more critical, becoming more sophisticated and addressing a large number of consumer needs and aspirations.

Characteristics of the Functional Foods and Nutraceuticals Revolution in the US

In no other country in the world is the development of functional food and nutraceuticals redefining the boundaries between what a food is, what a drug is, and what a dietary supplement is, and how all these are being used to maintain health, as in the US.

However, while US manufacturers are keen to bring about a functional foods and nutraceutical revolution – functional food has been one of the top US food industry research and development priorities for a number of years – they seem as yet uncertain how to do this. As this revolution unfolds, it is producing some unusual marketing and regulatory twists. For example Nabisco, a long-established food company with sales of around US$9 billion in 1999, launched its first ever dietary supplement product Knox *NutraJoint* in 1997. On the other hand Ross Products, a division of health care giant Abbott Laboratories with sales of US$11 billion in 1999, took its medical food Ensure, used in the dietary care of hospital patients, into mainstream consumer food markets. Other companies are also jumping regulatory hurdles that, if successful, could change the way Americans look at drugs, medical foods, dietary supplements and conventional foods in the future.

A key characteristic of the American market is the almost absolute focus on, and product-targeting of, disease and its prevention. Although often wrapped up in notions of 'wellness', there is an obsession with using food to target disease. An underlying theme, often evoked as the rationale for functional foods and nutraceuticals, is the aim of reducing the staggering health care costs in the US. In 1999 the US spent a trillion dollars on health care – half of the entire global expenditure on health (Bloom, 1999). At the same time, the American population, at one level, has never been in better health. For example, life expectancy has risen from 47 years in 1900 to 78 years in 1995; in the past two decades deaths from heart attacks and strokes have dropped by 30 to 50 per cent. Yet, certain sections of the population have never been more health-conscious – hence part of the drive to provide health solutions through functional foods and nutraceuticals.

But It's All a Word Game

The functional food market is characterized by a sophisticated regulatory 'word game' as regulators and industry grapple with products that seem to treat, cure, mitigate or prevent disease, yet are sold to the consumer as food or dietary supplements. Closely related to this is the development of health claims and labelling issues. Complications arise, because these terms – including drug, food, medical food and dietary supplement – all have strict legal definitions in the US, but there is no legal definition for functional foods or nutraceuticals.[1]

Until recently, dietary supplements, and regulations pertaining to them, were driving the market for nutraceuticals. Therefore the market consisted mainly of pills and capsules rather than foods. This was stimulated by regulations passed in 1994 that allowed statements of nutritional support, especially the so-called structure–function claims (see below) for dietary supplements without prior approval from the FDA. Although some legal experts saw no reason why structure–function claims could not in principle be used on foods, the FDA had taken a hard line against companies that tried to promote foods using such claims and sold as dietary supplements.

Regulatory Issues Governing Functional Food and Nutraceuticals

The functional food word game revolves around what constitutes a food or a drug, with all the regulatory nuances and permutations associated with this. It is a game for experts, and the purpose here is to offer something of a beginners guide to draw attention to key issues and developments which are shaping the market for nutraceuticals and functional food in the US, not to offer regulatory advice!

Food industry analysts sometimes cite regulation as holding back the functional food and nutraceuticals markets. Yet there are growing examples of products crossing regulatory boundaries and of companies not only pushing the spirit, but also the letter, of the law to drive forward this evolving dietary revolution, as well as many examples of companies successfully promoting health benefits within existing regulatory regimes.

Manufacturers who want to convey a product's benefits in labelling, and health and disease-related information have a number of options, but the most commonly used are using an FDA approved health claim or selling the product as a dietary supplement. But this is not the only route to market. For example, functional foods could be foods for special dietary or medical use, or foods could simply make nutrient content claims, or provide truthful and non-misleading information about the 'physiological' effect of a food. In this chapter we focus on developments using FDA approved health claims.

Ground-breaking Food Regulations Set the Scene for Functional Food

Two pieces of legislation in the 1990s set the scene for functional food and nutraceutical developments in the US – the NLEA implemented in 1994, and the DSHEA, which came into effect from March 1999. The NLEA is in many respects a triumph for the healthy-eating revolution in that it institutionalized many of the concepts of diet and health that form the core of healthy eating.

The NLEA evolved from the chaos caused by the healthy-eating revolution and the thousands of unregulated health claims being made by food processors in the wake of Kellogg's market intervention in 1984. It is not, therefore, forged in the spirit of the functional foods revolution, but as we shall see, it is now being used to this end through petitions for food-specific health claims.

The NLEA and Health Claims

In 1994, the US became the first nation to require that all processed foods disclose full nutritional information. Before the NLEA, nutrition labelling was only required if manufacturers made a nutrition claim about a product. Under this system, around 60 per cent of food labels disclosed nutrition information with often only minimal data on calories, protein, fat, carbohydrates and sodium contents, and in inconsistent formats (Silverglade, 1997). Mandatory nutritional labelling was the cornerstone of the NLEA, which is the most comprehensive revision of food labelling laws since the Federal Food, Drug and Cosmetic Act was passed in 1938. The goal of the NLEA, and its implementing regulations, was to 'bring a sense of order to the understanding of terms used when describing characterizations of food products'. The new law set comprehensive standards for nutrition claims such as light, low fat, and healthy. The NLEA also set up a regulatory framework for the approval of health claims such as 'Diets low in fat and cholesterol may help reduce the risk of heart disease'.

The major impetus for the enactment of the NLEA grew out of the general concern over the relationship between diet and disease, particularly heart disease and cancer, which are the two leading causes of death in the US (diet has been implicated in six of the ten leading causes of death in the US: heart disease, cancer, stroke, diabetes, atherosclerosis and liver disease).

The NLEA has three basic objectives (Hasler et al, 1995):

1 To encourage product innovations through the development and marketing of nutritionally improved food;
2 To eliminate consumer confusion by establishing definitions for nutrient content claims;

3 To make available nutrition information that can assist consumers in selecting food that can lead to healthier diets.

The NLEA now requires that practically all food products provide a 'Nutrition-Facts' label disclosing total fat, saturated fat, cholesterol, sodium, total carbohydrate, fibre, sugar, protein, vitamin A, vitamin C, calcium and iron content. This information must be listed for a customary portion of the food. The FDA has defined customary portions for more than 130 categories of food. One of the most novel features of the law is a requirement that the label should also disclose how much a customary portion of the food contributes to the total amount of these nutrients that an average American should consume per day in order to stay within official guidelines for healthy eating. As a result, in theory, consumers can now tell at a glance whether a food is high or low in particular nutrients, such as saturated fat or fibre.

A primary goal of the NLEA is also to protect consumers from unfounded health claims. This is reflected in the two rigorous standards now required for permitting health claims on food labels:

1 The health claim must be supported by the totality of the publicly available scientific evidence;
2 There must be significant scientific agreement among qualified experts that this support exists (Hasler et al, 1995). Although the FDA contends it will 'not require that health claims be supported by a full consensus among scientists', the issue of what constitutes significant scientific agreement is still the subject of much debate.

For functional food, the concern is that if standards for health claims are too high, few companies will be willing to accumulate the evidence needed to meet them.

The Keystone National Policy Dialogue

The passage of the Nutrition Labelling and Education Act of 1990 marked the culmination of considerable debate and controversy. Even after the enactment of NLEA, however, disagreements about aspects of the law and the rationale behind it remained, as did concerns about its implementation. A response to these disagreements was the setting up of a Keystone National Policy Dialogue on Food, Nutrition, and Health. The Keystone Dialogue brought together representatives from different interests to discuss and possibly reconcile such disagreements. The Keystone final report was published in March 1996, but seems to have had minimal impact on policy formation. However, it is interesting to note the outcome of the dialogue on health claims. The report says that the controversy over health claims stemmed from three general sources:

1 Fundamental differences in philosophy among dialogue participants about the appropriate role that government should play in regulating dissemination of information about the nutritional benefits of food products by the food industry;
2 A lack of understanding about how the FDA was and would be implementing NLEA, such as the application of the significant scientific agreement standard in determining the validity of a diet-disease relationship proposed for authorization as a health claim. Some people believe that the significant scientific agreement standard is essential for establishing a high level of confidence in the diet-disease relationships stated by health claims. Others believe that the standard is too strict to allow authorization of a sufficient number of health claims;
3 Lack of knowledge and understanding of the regulatory process used to implement NLEA, as demonstrated by the various perspectives concerning the FDA's review of the relationships between food substances and disease or health-related conditions mandated for evaluation in the law.

New Precedents for Health Claims

A number of generic health claims for broad categories of foods are currently approved under the NLEA. According to the system introduced by the NLEA, a health claim is any claim on the package label or other labelling of a food that characterizes the relationship of any nutrient or other substance in the food to a disease or health-related condition. Such claims are generic, not product-specific, and can therefore be used by any manufacturer whose product meets the required criteria.

To qualify for labelling with a health claim, foods must contain:

- A nutrient (such as calcium) whose consumption at a specified level as part of an appropriate diet will have a positive effect on the risk of disease; or
- A nutrient of concern (such as fat) below a specified level.

The NLEA initially detailed where claims were allowed in the following relationships:

- Calcium and a reduced risk of osteoporosis;
- Sodium and an increased risk of hypertension;
- Dietary saturated fat and cholesterol and an increased risk of coronary heart disease;
- Fibre-containing grain products, fruits and vegetables and a reduced risk of cancer;
- Fruits, vegetables and grain products that contain fibre, particularly soluble fibre, and a reduced risk of coronary heart disease;
- Fruits and vegetables and a reduced risk of cancer.

For each of these claims, the regulations provide strict criteria regarding minimum amounts of certain protective components and maximum amounts of risk-related components. At the same time the FDA denied claims for fibre and cancer; fibre and cardiovascular disease; antioxidant vitamins and cancer; Omega–3 fatty acids and heart disease; zinc and immune function; and folic acid and neural tube defects.

In 1994, the FDA reconsidered some of these and allowed health claims relating to folic acid consumption for the prevention of neural tube defects and health claims relating to an association between dietary fat and cancer, having concluded that diets low in fat may reduce the risk of some cancers. In August 1996, the FDA issued a final rule providing for a health claim for sugar alcohols and non-promotion of dental caries. The health claim may be used with eligible foods and dietary supplements.

Selected examples of FDA model health claims

- Calcium and osteoporosis. Regular exercise and a healthy diet with enough calcium helps teen and young adult white and Asian women maintain good bone health and may reduce their high risk of osteoporosis;
- For foods with calcium contents greater than 40 per cent of the reference daily intake (RDI) per reference amount. Regular exercise and a healthy diet with enough calcium helps teen and young adult white and Asian women maintain good bone health and may reduce their risk of osteoporosis later in life. Adequate calcium intake is important, but daily intakes above 2000 mg are not likely to provide any additional benefit;
- Fat and heart disease. While many factors affect heart disease, diets low in saturated fat and cholesterol may reduce the risk of this disease.

- Development of heart disease depends upon many factors, but its risk may be reduced by diets low in saturated fat and cholesterol and healthy lifestyles;
- Sodium and hypertension. Diets low in sodium may reduce the risk of high blood pressure, a disease associated with many factors;
- Dietary fibre and heart disease. Diets low in saturated fat and cholesterol and rich in fruits, vegetables and grain products that contain some types of dietary fibre, particularly soluble fibre, may reduce the risk of heart disease, a disease associated with many factors;
- Fruits and vegetables and cancer. Low-fat diets rich in fruits and vegetables (foods that are low in fat and that may contain dietary fibre, vitamin A, and vitamin C) may reduce the risk of some types of cancer, a disease associated with many factors. Broccoli is high in vitamins A and C, and it is a good source of dietary fibre.

Leveraging hidden nutritional assets

The US is taking the functional foods revolution in a new direction. Here we see one of the most important recent trends in functional foods, that is the way companies are leveraging their hidden nutritional assets. More and more traditional foodstuffs, such as oats, whole grains, tomatoes, soy, and berries are becoming the powerhouses of the functional foods revolution as companies use and skilfully market, modern nutritional science about the health benefits of everyday foodstuffs to consumers. Many of these companies, notably Quaker Oats, General Mills, The Kellogg Company, and Protein Technologies International, are using health claim regulation as part of their marketing armoury.

Of particular importance, in the context of leveraging hidden nutritional assets, has been the approval by the FDA of a number of food-specific health claims resulting from petitions to the FDA. The first ever product-specific health claim approval using NLEA regulation was in 1997, for oats and heart health (based on the cholesterol lowering properties of the soluble fibre – beta glucans – found in oats), following a petition from the Quaker Oats Company. The approval was later extended to include psyillium husks and heart health following a petition from The Kellogg Company.

Also important was the approval in July 1999 of a 'whole grain' claim in relation to heart health following a petition from General Mills, although this claim was achieved using a different regulatory approach – through the Food and Drug Administration Modernization Act of 1997 (FDAMA). More recently, on 26 October 1999, the FDA approved a 'heart health' claim (again relating to lowering cholesterol) for soy protein following a petition, again using NLEA regulation, from Protein Technologies International, part of the DuPont group, a major player in the soy market. Collectively, these developments move the US word game into a different league and the following section looks at these in more detail.

Quaker Oats Sets the Precedent for Food-Specific Health Claims

The first health claim for a specific food product was the result of a petition from Quaker Oats first filed in March 1995. FDA officials, acting in response to the petition, issued a final rule in 1997 allowing Quaker Oats and other food companies to claim that soluble fibre in oatmeal, oat bran and other whole-oat products may reduce the risk of heart disease. This was an unprecedented decision since until then the FDA had only allowed the generic health claims for broad categories of foods, as described above. This was also the first time that the FDA had acted in response to a petition from an individual manufacturer. However, the new ruling requires product labels to spell out to consumers that soluble fibre helps reduce the risk of heart disease only if it is eaten as part of a diet low in saturated fats and cholesterol.

At the time Steven Ink, the director of research for the company was reported as saying:

> *'The oatmeal health claim sets a high scientific standard for the rest of industry in proposing food-specific claims. Few foods can make health claims supported by more than 30 years of research and over 37 clinical trials worldwide...For so many years, consumers have been getting messages about what foods not to eat...This FDA announcement is about a positive message, namely, eat oatmeal and it will help you reduce your risk of heart disease.'* (for detailed case study of the Quaker Oats petition see Paul et al, 1999)

Cholesterol–Lowering Effects

In allowing this health claim, the FDA concluded that the beta-glucan soluble fibre of whole oats is the primary component responsible for the total and LDL blood cholesterol-lowering effects. To qualify for the FDA claim, products must contain 0.75 grams of soluble fibre (beta glucan) per serving. The amount of soluble fibre needed for an effect on cholesterol levels is about 3 grams per day (meaning a consumer would need to take four servings a day to achieve a cholesterol-lowering effect).

Below are examples of how the allowed health claim may be used:

> *'Soluble fibre from foods such as oat bran, as part of a diet low in saturated fat and cholesterol, may reduce the risk of heart disease.'*

or

> *'Diets low in saturated fat and cholesterol that include soluble fibre from oatmeal may reduce the risk of heart disease.'*

The words 'Diets low in saturated fat and cholesterol' must be included in any such health claim because the FDA concluded, after reviewing comments, that consumers might otherwise be misled into thinking that eating a diet high in oats is all that is necessary to reduce the risk of heart disease.

Kellogg petitions for Psyllium health claim

Following the oats claim, the FDA next approved a claim on 18 February 1998 in response to a petition from Kellogg (first published in the federal register on 21 May 1997) allowing a food-specific health claim linking soluble fibre from psyllium husks to reduced heart disease risk. The FDA final rule states:

> *'...the agency is persuaded by the evidence it has reviewed in this rulemaking that the consumption of pysllium seed husk, as part of a low saturated fat and cholesterol diet, can be a prudent public health measure to assist in the national policy of promoting eating patterns that will help in achieving or maintain-*

ing desirable blood cholesterol levels in the general population.'

In 1999 Kellogg launched a range of psyllium-based products under the brand name Ensemble. These products were a 'complete family' of functional foods from pasta to cereals, but they failed in their test market and were withdrawn from the market in the same year.

A 'Whole Grains' First

The FDA health claim approval secured by General Mills for whole grain using FDAMA legislation is again a first. FDAMA allows for health claims without FDA pre-approval if such claims constitute statements made by authoritative government bodies such as the Centre for Disease Control, the National Academy of Sciences or the National Institutes of Health. General Mills' notification to the FDA relied upon the executive summary of a National Academy of Sciences report, *Diet and Health: Implications for Reducing Chronic Disease*, which the FDA accepted as an 'authoritative statement'. So from 8 July 1999 manufacturers could use the following claim on the label and in the labelling of food products that contain 51 per cent or more whole grain ingredients:

'Diets rich in whole grain foods and other plant foods and low in total fat, saturated fat, and cholesterol, may help reduce the risk of heart disease and certain cancers.'

Like other FDA-approved health claims, the claim is generic in the sense that anyone can make the claim providing they meet the product criteria, but General Mills is proving to be the master player at leveraging commercial benefits from nutrition marketing built on these FDA-approved food-specific health claims. In December 1999 General Mills just nudged ahead of The Kellogg Company as the top US cereal company, when measured by value shares (Kellogg still had the largest volume share), and in the half year from June 1999 to January 2000 General Mills achieved total sales of US$3.39 billion and net earnings, after tax, of US$352 million.

General Mills' flagship nutrition brands are products such as Cheerios, Wheaties and Whole Grain Total – brands like these account for a third of their cereal volume. Following the earlier FDA approval for a health claim for heart health in relation to oats consumption in

1998, the company has used this brilliantly in the marketing of its oat-based breakfast cereal Cheerios. The irony is that the FDA approval for the claim was the result of a petition by cereal market competitor Quaker Oats, yet it could be argued that General Mills has gained most commercially from the claim through their use of it in marketing and communication strategies. In 1999, according to General Mills, Cheerios outperformed the cereal category for the second year running, posting an 11 per cent volume gain, and in the first half of 2000 was six per cent ahead again. Now using the whole grain concept General Mills is again making the claim pay in the marketplace. In the first 11 weeks of advertising and promotion based on the whole-grain claim, its whole grain cereals gained a seven per cent market share. What is remarkable is that the company is using nutrition marketing to give new life to Cheerios – its baby-boomer brand which in 1999 was already 57 years old!

General Mills is now taking the whole grain concept further, using it to stake out a nutritional marketing 'high ground' around whole grain consumption. The whole grain concept is also being extended worldwide. For example, through its products made by its European alliance Cereal Partners (a 50/50 joint venture between General Mills and Nestlé), the same whole grain logo, as used in the US, is appearing on the appropriate UK cereals.

Health-Claim Ruling Marks Climax of the 'Year of Soy'

In another landmark move, on 26 October 1999, the FDA authorized the use of a health claim, linking consumption of soy protein to reduced risk of heart disease, on food labels. In the months preceding the FDA's ruling, expectations mounted as companies jostled to position themselves to profit from the expected boom in sales of soy products, prompting 1999 to be labelled by many 'the year of soy' (in Europe, of course, 1999 was a very different 'year of soy' when there was a massive backlash against GMOs – genetically modified organisms – particularly GM soy).

The FDA's ruling came in response to a petition filed in 1998 by Protein Technologies International, part of the DuPont group and one of the world's largest suppliers of soy ingredients. Scientific evidence supports a strong connection between consumption of soy protein and a reduction in levels of blood cholesterol and Protein Technologies International sought to capitalize on this by enabling companies making soy-based foods to use a so-called 'heart-healthy' claim on their products. There is a potentially huge market for heart-healthy foods in the US based on soy, with the US being one of the world's major soy producers. However, currently, the majority of soy production goes for animal feed and in 1999 only about three per cent of production went for human consumption.

The FDA said that the research on soy showed that 25g of soy protein a day is needed to get a clinically significant effect on total cholesterol levels and agreed with PTI's proposal that a single serving of a soy protein-containing product should provide 25 per cent of this total (assuming four servings a day). In making its decision, the FDA conducted a comprehensive review of 41 human studies submitted in the PTI petition. Of these the agency gave weight to 14 trials, which included subjects representative of the US population; reported saturated fat and cholesterol intakes; and did not have 'design problems' such as small sample size and lack of a placebo, for example.

The FDA allows the heart-health claim for foods containing a minimum of 6.25g of soy protein per serving. Soy now joins oat beta glucan and psyllium as foods allowed to state that they may reduce the risk of heart disease. Claims can take one of two permitted forms. These are:

1 'Twenty-five grams of soy protein a day, as part of a diet low in saturated fat and cholesterol, may reduce the risk of heart disease. A serving of (name of food) supplies Xg of soy protein';
2 'Diets low in saturated fat and cholesterol that include 25g of soy protein a day may reduce the risk of heart disease. One serving of (name of food) provides Xg of soy protein.'

The FDA has also granted an exemption to the low fat requirement for foods derived from whole soy beans such as soy milk. Under the terms of the ruling each serving must contain no more than three grams of fat, 20mg of cholesterol and 480mg of sodium.

Companies such as Protein Technologies International, Archer Daniels Midland and Central Soya and other major food processors are working to ensure that soy ingredients can be formulated into a wide range of foods such as bread, breakfast muffins, breakfast cereals and pasta.

Traditional Foodstuffs Becoming Functional Foods

In March 2000 The Kellogg Company, the radical of nutrition marketing, looked to 'push the envelope' even further. Using a structure–function claim for folic acid, Kellogg announced it will promote the heart-healthy benefits of the nutrient. This makes it the first major food company to communicate, in this case on adult cereal products, that 'adequate intakes of folic acid, vitamins B6 and B12 may promote a healthy cardiovascular system'. Consumers are able to identify what the company calls 'newly fortified' cereals through a red star and a blue banner that says 'Good News for Your Heart – 100% Daily Value Folic Acid, B6, B12'.

These developments, and others based on traditional foodstuffs (see Chapter 10 for further examples) see the functional food concept in the US entering a new era and being driven by some of the world's largest and most powerful food corporations. These food-specific health claims for traditional foodstuffs are being so skilfully appropriated by generally the larger food companies on a host of products, that the single, product-specific health claim (such as on a spread like Benecol) is in danger of being eclipsed. Developments in traditional foodstuffs, while using new scientific understanding of the health benefits of dietary components, are fast making all food 'functional'.

The US Market for 'Nutrition' Products is Booming

The developments in these traditional foodstuffs and FDA-approved health claims, however, should be seen as running parallel to a booming natural products and dietary supplements industry in the US during the last part of the 1990s. The American market for nutrition products, such as dietary supplements, has seen dramatic growth during the 1990s (apart from 1998 when the market temporarily 'crashed') and has been a key focus for nutraceutical and functional food product development. A particular explosion in herbals has helped fuel the general boom. According to the US industry newsletter *Nutrition Business Journal* (NBJ), nutraceuticals – which are broadly defined to include a range of nutrition products, functional foods and certain other health-beneficial foods – was worth US$86 billion in 1997, broken down as follows: market standard foods (35 per cent); 'lesser-evil foods'(29 per cent); functional foods (17 per cent); dietary supplements (12 per cent); natural foods (9 per cent) and medical foods (less than one per cent).

In the case of dietary supplements, Table 7.1 shows that the market value increased by just over 85 per cent between 1994 and 2000, with growth in all categories. Eighteen companies with sales greater than US$100 million accounted for nearly two-thirds of total sales in this market.

Table 7.1 *The US Nutrition Industry (1994–2000) (US$million)*

Products	*1994*	*1995*	*1996*	*1997*	*1998*	*1999*	*2000*
Vitamins	3870	4250	4720	5190	5480	5790	6090
Herbals	2020	2470	2990	3530	3980	4400	4800
Sports nutrition	900	990	1070	1270	1420	1530	1620
Minerals	690	800	890	1,050	1140	1210	1280
Meal supplements	450	560	590	640	660	680	700
Specialty/other	670	750	920	960	1180	1330	1480
Supplements total	8610	9820	11,180	12,640	13,860	14,940	15,970

Source: Nutrition Business Journal, 1999

Table 7.2 *Top 20 US Supplement Manufacturing/Marketing Companies in 1998*

Company	*Sales (US$million)*
Leiner Health Products	580
American Home Products	540
Rexall Sundown	531
Pharmative (Otsuka Pharm)	410
NBTY	350
TwinLab Corporation	321
General Nutrition Products Inc	320
Weider Nutrition Group	290
Perrigo	178
Bristol-Myers Squibb (Mead Johnson)	160
Bayer Corporation	150
Experimental & Appl. Sciences (EAS)	135
Murdock Madaus Schwabe (N Way/B&T)	130
Powerfoods Inc	115
IVC Industries Inc	113
Country Life	110
MET-Rx USA	108
Nutraceutical International	105
Amrion Inc (Whole Foods Market)	84
Balance Bar Co	82

Source: Nutrition Business Journal, 1999

According to the NBJ, twenty-five million consumers in the US are regular purchasers of dietary supplements, spending between US$20 and 50 per month. These consumers account for 70 per cent of all supplement purchases (other estimates say more than 100 million people in the US take dietary supplements daily).

DSHEA Allows Statements of Nutritional Support[2]

The booming market for dietary supplements and related nutrition products has driven US interest in the potential for functional food and nutraceuticals. In fact, during the mid-1990s the majority of nutraceutical activity has been in dietary supplements rather than foods.

The market has been driven in part by the Dietary Supplement Health and Education Act of 1994. DSHEA has been seen as opening up the market for functional food and nutraceuticals, particularly in relation to the use of health claims on products. In this respect, as well as creating a new category of food by specifically defining dietary supplements as foods, DSHEA allows dietary-supplement labelling to bear certain statements of 'nutritional support' (see Appendix 3 for the main provisions of DSHEA).

This includes structure–function claims, as they are known, which differ from health claims defined under the NLEA and do not need prior FDA approval. All a company has to do is send the FDA a notification letter that it is making a claim. It is up to the FDA to challenge claims made under the DSHEA. This Act does not allow a manufacturer to make a drug claim for a product – that is, a statement claiming to 'diagnose, mitigate, treat, cure, or prevent a specific disease'. However, under the Act, a dietary supplement may no longer be deemed a drug solely because it includes a statement of nutritional support. Table 7.3 gives some examples of structure–function claims from notification letters to the FDA in 1997. By the beginning of 1998 more than 1000 structure–function claims notification letters had been sent to the FDA, rising to well over 3000 in 2000. If a structure–function claim is made, products also have to carry a 'disclaimer' on product labelling that they have not been evaluated by the FDA and are not intended to 'diagnose, treat, cure or prevent disease'. Dietary supplements are mainly sold as tablets, capsules and powders, and not as food-like products.

In short, the DSHEA creates a broad, new definition of dietary supplement products, in theory moderates the regulatory burdens for the use of dietary ingredients in dietary supplements, and permits additional use of promotional information about the benefits of dietary supplements, both through the use of statements of nutritional support on labels and in other labelling, and by enabling sellers to refer customers to reprints of books, articles, and other publications that provide health-related information. DSHEA has been seen as offering substantial opportunities for the promotion of health-related benefits that might not otherwise be available without compliance with drug standards or approval of a health-claim regulation.

Table 7.3 ***Examples of Structure–Function Claims sent to the FDA***

Company and date	*Product/ingredient*	*Claims*
Advanced Plant Pharmaceuticals (January 1997)	Lo-Chol methachol	'Scientifically proven to lower cholesterol levels, shown in numerous clinical studies to not only lower total cholesterol and triglycerides, but also increase HDL ("good" cholesterol) and balance the proportion of HDL to LDL'
NutraGenics (January 1997)	Evolve with clearesterol	'Helps lower cholesterol'
Pershing Products (January 1997)	LDL-Lite	'Reduce your cholesterol up to 15 per cent in four weeks'
Phoenician Herbals (January 1997)	Detox tea	'Red clover – acts as a blood purifier, also a remedy for nervous exhaustion by calming the nerves. Liquorice – helps protect the body's detoxification plant, our liver, from disease. Burdock – cleanses and eliminates long-term impurities from the blood. Has the ability to neutralize most poisons, relieving both kidneys and lymphatic systems. Dandelion – regarded by herbalists as one of the best herbs for building up the blood and for helping with anemia. Yellow dock – is used to combat liver disease and gall bladder problems. *Cascara sagrada* – sparks up a sluggish colon that is chronically constipated by improving the flow or secretions from the stomach, liver and pancreas'
Rexall Sundown (January 1997)	Sundown vitamins Folic acid	'Helps nutritionally in maintaining cardiovascular function and a healthy circulatory system by reducing the amount of homocysteine levels in the blood. Elevated homocysteine levels are associated with plaque buildup in the arteries and reduced circulatory function'

Company and date	*Product/ingredient*	*Claims*
Shaklee (January 1997)	Premium garlic	'A variety of studies have shown that daily intake of garlic may help lower blood cholesterol levels'
Sinomedical Development (July 1997)	Gu Kany Ye	'Researchers in laboratories and clinics have shown that consumption of this product can improve blood circulation and bone growth of humans'
NaturalLife (May 1997)	Chromium picolinate	'Helps maintain normal metabolism. Chromium is an essential part of the Glucose Tolerance Factor (GTF) molecule. GTF is an important co-factor for insulin in the regulation of blood sugar which is necessary for proper metabolism'
Yerba Prima (June 1997)	St John's Wort tablets	'Promotes emotional well-being. It has been shown to balance moods and emotional function under stress and to support a general feeling of well-being'

The Act therefore provides a company with some significant opportunities for devising promotions which will convey information about health-related benefits, but will not either trigger drug status or require FDA approval of a health-claim regulation. In practice, however, DSHEA has opened up a whole new area of regulatory confusion and taken the US word game to new heights (and lows) of gamesmanship.

Dietary Supplement or Conventional Food?

A good example of regulatory confusion was the way in which, using DSHEA regulation, a number of food products have been launched as dietary supplements. For example a breakfast cereal *GinkgOs* ('For Sharp Thinking') was launched as a herbal dietary supplement by New Morning, Acton, MA in 1998, but is no longer on the market. Together with the required regulatory disclaimer 'this product is not intended to diagnose, treat, cure or prevent disease', its packaging made the following claims: 'Improves Blood Flow to the Brain', 'To Sustain Memory', 'Antioxidant', 'Wheat Free' and 'Low Sodium'.

March 1998 saw the launch of what was claimed to be the first milk-based range of products to be sold as a dietary supplement. *Basics Plus*, from Lifeway Foods is a fermented dairy drink kefir but contains *Proventra*, the brand name for natural immune components prepared by the biopharmaceutical company GalaGen, from the colostrum of dairy cows. In 1999, France's leading dairy company Danone announced it had taken a 20 per cent stake in Lifeway Foods.

As we discussed in Chapter 2 on the attempt by both McNeil and Unilever to launch cholesterol-lowering spreads in the US as dietary supplements, the FDA is starting to take a tough stance on what it sees as the distinction between 'food' and 'dietary supplement'. In this example the FDA persuaded both McNeil and Unilever from marketing their products as dietary supplements arguing they were 'food' and should be marketed as much (however, the FDA were apparently happy for both companies to use 'structure-function' claims on package labelling).

In August 1999, the dairy company Dannon introduced a functional dairy product called Actimel into the US (see Chapter 9 on the European dairy industry for full description of Actimel), again as a dietary supplement, but the FDA also challenged this, saying the product was a food and not a dietary supplement. At this stage it would appear the dietary supplement route for functional food products is being closed, but of course it is still open to challenge.

Drug or Dietary Supplement?

Equally as far-reaching is the blurring of distinctions between drugs and dietary supplements. This is illustrated by the product Cholestin, made from *hong qu*, a red-rice yeast marketed as a dietary supplement making a cholesterol-lowering claim. Cholestin is sold by Pharmanex, a company formed in the wake of DSHEA to manufacture and market dietary supplements under the new law. However, Merck the drug company marketing Mevacor (lovastatin), regulated as a drug, challenged the fact that Cholestin could be sold as a dietary supplement, saying that it was in effect a drug. Merck argued that the red-rice yeast is a 'generic lovastatin product that was designed as, and is being aggressively marketed as, a generic alternative to Mevacor'. Pharmanex argued that Cholestin is marketed squarely within the DSHEA as a dietary supplement.

The FDA took Pharmanex to court arguing that Cholestin was in fact a drug. In the subsequent court case held in Utah in 1999, Pharmanex won the argument that their product was in fact a dietary supplement. Then on 21 July 2000, the Court of Appeals rejected the rationale of the District Court in Utah. However, while an apparent victory for the FDA, the Court of Appeals did not rule that Cholestin was necessarily a drug and the case is back in the courts for further clarification. But as McNamara (2000) says:

> *'In the meantime, FDA is likely to be encouraged by this ruling by the Court of Appeals, and the agency may become more confident in challenging the dietary supplement status of products that contain a substance that is also present as an active ingredient in an approved new product (assuming the substance was not marketed as a dietary supplement or as a food before the new drug application was approved.'*

A Continuing Regulatory Saga

Slowly, the US is moving towards a regulatory regime that allows greater opportunities to provide explicit health-related information in labelling for food and other products. Yet it is an evolving regulatory environment. For example, in the Federal Register of 6 January 2000, the FDA published new final regulations (in relation to dietary supplements) that specify whether particular types of claims will be deemed by the Agency to be acceptable disease claims (that is, not acceptable structure–function claims). The FDA has published a long list of permissible structure-function claims and impermissible disease claims. But the unfolding of the DSHEA 'word game' still has a long way to run and there is much that is still subject to interpretation and, it should be assumed, testing in the courts.

Another important case with the potential for far-reaching implications on health claims in the US, is that of Pearson, held in the District of Columbia Circuit in 1999. In a unanimous three-judge opinion, it was ruled that the FDA had acted improperly in denying four particular 'health claims'. The Court did not authorize the use of the claims, but instead sent the matter back to the FDA for further consideration. Among a number of questions the Court asked the FDA to consider was the possibility of approving 'health claims' that incorporate qualified representations or 'disclaimers'. When this book went to press the implementation of the Pearson decision was still pending, but many in the food industry, and regulatory experts, see the Pearson case as having the utmost importance.

Box 7.1 Market Case Study: Campbell's IQ Range Dims Functional Food Aspirations

One of the most significant business developments in functional foods during 1997–98 that caught the imagination and attention of the US food industry was a range of functional food products called Intelligent Quisine (IQ) introduced by the Campbell Soup Co. Intelligent Quisine was a range of 'clinically-proven' mail-delivered meals designed for people with cardiovascular problems, type II diabetes and other health concerns.

IQ was the first and only meal programme clinically proven to reduce high blood cholesterol, high blood pressure and high blood sugar in individuals with one or more of these conditions. In other words, it was the first major food venture to be modelled after a pharmaceutical business plan. One of the many interesting features of the IQ programme is that even though it made strong claims, for example, to reduce high blood cholesterol, it was classified by the FDA as a conventional food and not a medical food, which was very favourable from Campbell's point of view.

Promotional materials show the high hopes the company once had for IQ, calling it their 'most significant' scientific and business commitment 'since the development of technology to make condensed soup'. Campbell is the world's largest soup maker, with nearly US$8 billion in annual sales. While plenty of food companies sell low-fat, low-salt frozen dinners in supermarkets, Campbell was trying to become the first food giant to establish a market for functional foods designed for people with specific health conditions.

After conducting three clinical trials with more than 800 subjects, in consultation with the American Heart Association and the American Diabetes Association, Campbell sent its sales representatives to doctors' offices, just like a drug company, and targeted the public in advertisements. Advertisements for IQ boasted that its frozen meals and snacks could reverse certain medical conditions such as high blood pressure and high blood sugar – one proclaimed 'I ate cheesecake and my cholesterol went down 15 points'. But analysts said IQ never came close to meeting the US$200 million annual target that Campbell predicted in September 1996.

Campbell took five years and reportedly invested US$30–50 million developing IQ. The products were test-marketed in Ohio at the end of 1997. Consumers choosing to participate in the IQ programme had to commit themselves to a four- or ten-week plan that included three meals and one snack with a choice of 41 specially formulated varieties of meals at a total cost of US$80 a week. In February 1998 the company reported that its trial was paying off with 40 per cent of participants remaining with the meal plan after its first phase, although there were no immediate plans to move the programme beyond Ohio.

However, in April 1998 the company announced it was 'pulling the plug' on the IQ range, saying the decision was related partly to the US$11.5 billion sale of Vlasic Foods International, part of the Campbell food empire. 'We took a hard look at the resources that would be necessary to continue the programme, and decided we didn't want to go forward with it', a spokesman said. He wouldn't discuss sales for IQ, adding only that the mail-order product line 'didn't meet Campbell's business expectations'.

It is a common belief in the US food industry that IQ was five years 'too early'. It has now been sold to a small company that is investigating ways of reintroducing the products – IQ is not yet dead and buried (see Chapter 10 for more examples of US functional food product developments).

Consumers at Risk from Functional Foods and Dietary Supplements?

As the market for functional foods and dietary supplements has developed it has drawn the attention of policy makers and, rather belatedly, of consumer advocacy organizations. In July 2000, the influential US General Accounting Office (GAO), the investigative arm of the US Congress, questioned the ability of existing US regulation to ensure the safety of functional food and dietary supplements and warned that current federal laws do little to protect consumers against inaccurate and misleading claims.

The General Accounting Office says consumers' health may be put at risk because the Food and Drug Administration lacks the resources to investigate potential adverse health effects from consuming dietary supplements and functional foods. In their report to Congress, *Improvements Needed in Overseeing the Safety of Dietary Supplements and 'Functional Foods'* (2000), the GAO recommends that Congress:

- Amend the Federal Food, Drug, and Cosmetic Act to require makers of functional foods to meet the same requirements that currently apply to dietary supplements; and
- Establishes an expert panel to re-examine the current approach to labelling, which distinguishes health claims from structure-function claims, to determine whether the intended distinctions can be made clear and meaningful to consumers, or failing this, to identify changes needed to improve consumers' understanding of health-related claims.

What the GAO Report Says

The GAO report found that FDA efforts and federal laws provide limited assurances of the safety of functional foods and dietary supplements.

They concluded that while the extent to which unsafe products reach consumers is unknown, three weaknesses in the regulatory system increase the likelihood of such occurrences:

1. Potentially unsafe products may reach consumers for a variety of reasons , including the lack of a clearly defined safety standard for new dietary ingredients in dietary supplements;
2. Some products do not have safety-related information on their labels, which could endanger some consumers. This occurs because FDA has not issued regulations or guidance on the information required;
3. FDA cannot effectively assess whether a functional food or dietary supplement is adversely affecting consumers' health because, among other things, it does not investigate most reports it receives of health problems potentially caused by these products.

The GAO also found that agencies' efforts and federal laws concerning health-related claims on product labels and in advertising provide limited assitance to consumers in making informed choices and do little to protect them against inaccurate and misleading claims.

The GAO acknowledges the 'blurring' between supplements and foods

The GAO report says that dietary supplement regulation in the US that allows supplements to exist in conventional food form, for example a nutrition bar, has blurred the boundary between foods and supplements. For example, during their review of the topic, GAO encountered several drinks, teas and other products, some produced by large companies with national distribution, that contain herbs – such as St John's Wort, gingko biloba, and echinacea – that appear to be functional foods, but have been 'incorrectly' marketed as dietary supplements. The GAO says:

> *'In order for these herbs to be legally marketed in food products, the producers would either have to determine that they are generally recognized as safe [GRAS status]*

for use in food or have them approved as food additives. FDA told us that it has not approved these herbs as food additives and further stated that the agency is unaware of any basis for concluding that these herbs are generally recognized as safe for use in food.'

Health Problems with Dietary Supplements Unknown?

The GAO says that the FDA may not know the extent of health problems associated with functional foods and dietary supplements and does not fully evaluate reported problems. Since 1993, the FDA has received 2797 reports of adverse effects, including 105 deaths, potentially associated with dietary supplements. Supplements containing ephedrine alkaloids, which are promoted for such effects as losing weight and increasing energy levels, have accounted for about 1173 reports of adverse effects, more than any other type of dietary supplement.

As of 29 February 2000 the FDA had not received any reports of health problems associated with functional foods, although the herbs most frequently associated with adverse reactions in supplements – ginseng, St John's Wort, echinacea, and ginkgo – have begun to appear in functional foods and beverages.

The GAO report suggests that adverse effects could be greater than the FDA figures reveal. They cite a 1999 nationwide consumer survey that found 12 per cent of all consumers who have used a herbal dietary supplement, about 11.9 million people, said that they had experienced an adverse reaction. Furthermore, the GAO was told that only three FDA staff are available part-time to investigate reports of health problems associated with dietary supplements.

Limited Consumer Protection Against Misleading Health Claims

The GAO report says that agencies' efforts and federal laws concerning health-related claims on product labels and in advertising provide limited assistance to consumers in making informed choices and do little to protect them against inaccurate and misleading claims. They cite three reasons for this:

1. FDA has not clearly established the nature and extent of the evidence companies need to adequately support structure–function claims and has taken no actions in court against questionable claims;
2. Federal agencies (for example the Federal Trade Commission and the FDA) operate under different statutes for regulating claims on product labels and in advertising, which has led to advertising claims that are not allowed on product labels;
3. Consumers may not understand the different purposes of health and structure–function claims. Consequently, they may incorrectly view structure–function claims as claims to reduce the risk of or treat a disease, rather than their more limited purpose of describing how consuming a product affects a structure or function of the body or a person's general well-being.

The GAO Conclusions

The GAO concludes that the FDA lacks an effective system to track and analyse instances of adverse effects and, until it has one, consumers face increased risks because the nature, magnitude, and significance of safety problems related to consuming dietary supplements and functional foods will remain unknown. The GAO made a number of recommendations to the Food and Drug Administration, namely:

1. Develop and promulgate regulations or other guidance for industry on the evidence needed to document the safety of new dietary ingredients in dietary supplements;
2. Clarify the boundary between conventional foods, including functional foods, and dietary supplements, particularly the circumstances under which dietary supplements may be marketed in food form;
3. Develop and promulgate regulations or other guidance for industry on the safety-related information required on labels for dietary supplements and functional foods;
4. Develop and enhance systems to record and analyse reports of health problems associated with functional foods and dietary supplements;
5. Develop and promulgate regulations or other guidance for industry on the evidence needed to support structure–function claims; and
6. Develop and implement a strategy for identifying and taking appropriate enforcement actions against companies marketing products with unsupported structure–function claims on their labels.

Consumer Advocacy Group Urges FDA to Halt Sale of 'Functional Foods' Containing 'Illegal' Ingredients

One of the US's highest profile consumer advocacy organizations, the Center for Science in the Public Interest (CSPI) asked the FDA in July 2000 to take regulatory action to halt the sale of 75 'so-called functional foods'. The CSPI, in 158 pages of written complaints about 75 products it had identified, said the products raise important health and safety issues and asked the FDA to order manufacturers to stop making false and misleading claims about them.

Bruce Silverglade, CSPI director of legal affairs said at the time: 'Food companies are spiking fruit drinks, breakfast cereals, and snack foods with illegal ingredients and then misleading consumers about their health benefits. It's shameful that respected companies are selling modern-day snake oil.'

The CSPI said that each of the companies named in the complaints markets products that are 'adulterated' because they contain ingredients that are not authorized by a 'food additive regulation' or 'prior sanction' and are not considered to be 'Generally Recognized as Safe'.

Products are 'Misbranded'

The CSPI also said: 'Each of these companies also markets products that are "misbranded" because they make label statements that are either false or misleading or make unapproved health or nutrient content claims.'

While acknowledging that not all functional foods are worthless, the CSPI said that problems with functional foods arise when:

- The added substance is not considered by the FDA to be safe for use in food;
- The purported benefit is based on flimsy evidence;
- There's only a trivial amount of the added substance in the food;
- The fortified food is loaded with fat, sodium or added sugars; and
- Consumers replace natural foods (like fruit and vegetables) that contain a wide variety of nutrients with manufactured foods that contain only one or a few added substances.

In their letter to the FDA urging them to take action against the products, the CSPI single out in particular herbals such as St John's Wort, kava kava, ginkgo and echinacea, and their use in products that are promoted for their added health benefits. In their FDA letter CSPI say:

> *'Manufacturers and distributors of functional foods containing these [herbal ingredients] and other ingredients are blatantly flouting the law that prohibits the use of substances that are neither approved as food additives nor GRAS. In addition, these products contain numerous outrageous claims on their labels.'*

Products that decades ago would have been considered to be 'snake oil' are reappearing today as 'functional foods'. Some of these claims are in direct violation of the Nutrition Labelling and Education Act. Others are simply false or misleading. In both cases, the claims render the products misbranded.

While the FDA has taken action against a few illegally marketed functional foods, those actions are woefully inadequate, given the increasing numbers of such products on store shelves. The FDA should promptly take regulatory action against all companies that are marketing products that contain unauthorized ingredients and make unlawful claims.'

CSPI: 'Over-the-top Dietary Puritanism'

The Grocery Manufacturers of America (GMA) said the CSPI call for a ban on the named functional products as a 'frenzied overreaction', dubbing the CSPI as 'food police'. GMA's vice president for scientific and regulatory policy, Stacey Zawel said at the time:

> *'CSPI is frankly engaging in over-the-top dietary Puritanism in its call for product bans. The FDA already has sufficient authority to take enforcement action against products it deems to be unsafe. Mainstream food companies fully support the high standard that all claims be scientifically substantiated, truthful and not misleading and that all products are safe. CSPI may not like it, but food companies have a constitutional right to make health-related information available to consumers.'*

Meanwhile the National Food Processors Association (NFPA), the voice of the US food processing industry on scientific and public policy issues, urged the CSPI to stop attacking functional foods as a category, and instead work with government and the food industry to ensure that consumers have access to important information on the health benefits of foods.

Dr Rhona Applebaum, executive vice president of scientific and regulatory affairs for the NFPA said:

> *'CSPI somehow misses the point that functional foods are foods. As such, they must meet existing strong regulatory requirements for safety and accuracy of label statements. There is no need for functional foods to be regulated differently than conventional foods; the Food and Drug Administration already has ample regulatory authority to oversee the safety of these foods and to ensure that claims on these products do not mislead consumers.'*

The Future

In a market report on functional foods in the US, published by *Decision Resources* in May 1998, author Rhonda Witwer summarizes what has happened to functional foods and the US food industry:

> *'Functional foods have made considerable progress over the past few years in capturing the attention of the entire food industry. Only three years ago, functional foods and nutraceuticals were considered to be on the periphery of the food industry. Selected nutrition experts, leading edge companies and some university groups (notably the University of Illinois's Functional Foods for Health Program) were arguing there was a solid scientific foundation behind functional foods and that they were not just a passing fad. Now, industry segments such as cereals, orange juice and other beverages, dairy and prepared meals have growing niche markets featuring health-oriented products containing bioactive ingredients.'*

Witwer reminds us that the functional food market is very new and is still unfolding. But even with this in mind, it is also quite remarkable to observe the scale of activity both at the market, policy and regulatory levels. There has been an almost constant stream of change and development over a very short period, notably from 1997 to 2000.

But as Witwer again points out, the food industry generally is not experienced in obtaining regulatory approval, even for food additives, let alone functional foods and nutraceuticals. The industry's ingredient suppliers provide them with additives and ingredients that are approved for food use. The regulatory skills within the industry typically deal with complicated labelling issues and good manufacturing practices, rather than with the technology and approval of food ingredients. The food industry as a whole has an aversion to high-risk research and development and prefers to introduce many new products using familiar technologies – knowing that only a few are likely to succeed. Furthermore, while the industry has a short-term profitability focus, functional food and nutraceutical development is a long-term effort exacting comparatively high costs.

However, despite these apparent difficulties, functional foods and nutraceuticals are being developed for many different markets in the US. Some are targeting particular medical conditions; others are targeting performance and endurance. Additional products are targeting disease prevention, including anti-ageing. Just how fast these markets will grow, and if they will ever become dominant, is under debate. Part of this growth will depend upon:

- How fast scientific research uncovers these diet, health and genetic connections;
- How fast the population learns about these connections; and
- How well functional foods and nutraceuticals anticipate and meet the complex needs of the population.

One major lesson from the US is how the effective use of the media plays a central role in a company's ability to succeed in functional food. Some of the most successful product developments have been based on the ability to leverage nutrition information and scientific findings in the aggressive media campaigns on the health benefits of products (see Chapter 10 for examples). In the mid-1990s the focus was on dietary supplements, but by 1999 the focus was changing to the marketing of traditional foodstuffs as functional food. Even though a number of food-specific health claims have now been approved by the FDA, it has been the way that particular companies have exploited these messages through communications and public-relations strategies – not by the on-pack claim itself – that has seen market success.

There is now a growing momentum to reposition traditional foodstuffs on the basis of their health-promoting properties, with the aggressive nutrition marketing of oats, soy, tomatoes, whole grains and certain juice-based products, and one of the biggest fortification trends is for calcium, hardly a new a health ingredient! Such generic nutrition marketing trends are threatening to eclipse single product nutrition marketing. Why should consumers purchase, for example, a single product offering to lower cholesterol, when they can get the same health benefit from a mass of 'generic' traditional foods such as oats, soy and whole grains?

Finally, in the area of food policy and regulation related to nutrition, health claims, foods and dietary supplements, there is unprecedented change and debate in the US. In these areas alone, the degree of activity compared to the rest of the world has been nothing short of remarkable. It has, perhaps unexpectedly, put the US at the cutting edge of food policy and is opening up the food economy to the full potential of functional foods. From a public health perspective, the biggest challenge of all is how the functional foods revolution can be delivered, and be of benefit, to the vast majority of its citizens.

Chapter 8

Is Finland the 'Silicon Valley' of European Functional Foods?

In Chapter 2 we described how the cholesterol-lowering margarine Benecol, invented by the Finnish company Raisio, had come to symbolize the concept and potential of functional food in the minds of many in the global food industry. But it is perhaps no accident that Benecol came out of Finland.

Finland, with a population of just over 5 million, is a remarkably innovative country in the context of functional food with the potential to develop a highly integrated functional food 'cluster' driving new product development. It is also possible, however, that Finland could lose its cutting edge in food and health developments unless action is taken to drive home the advantage of this unique, but in our view underdeveloped and under-performing, cluster. Finland appears to be on the crest of a functional food wave, but the prospect of Finland developing a 'functional foods Nokia' or 'Silicon Valley' is currently little more than wishful thinking.

According to Michael Porter, the grandee of cluster theory, clusters are geographic concentrations of interconnected companies and institutions in a particular field (Porter, 1990, 1998a and 1998b). Their advantage in the global economy, in Porter's view, is that they create enduring competitive advantage through local knowledge, relationships and motivations. A cluster allows each member to benefit as if it had greater scale or had joined with others formally, without requiring it to sacrifice its flexibility.

Finland's Ministry of Trade and Industry describes the structure of Finnish industry in general as one of clusters, in which different branches are closely linked with each other. The biggest clusters are in forestry and forest products, metal products, electronics, energy and chemicals. Although on a much smaller scale, it is clear from our research that an informal cluster has been established around food and health comprising universities, government-supported organizations, companies and individuals (Figure 8.1). However, within Finland, such a cluster is not yet formally recognized in economic terms and as a result the country may be missing a unique competitive advantage.

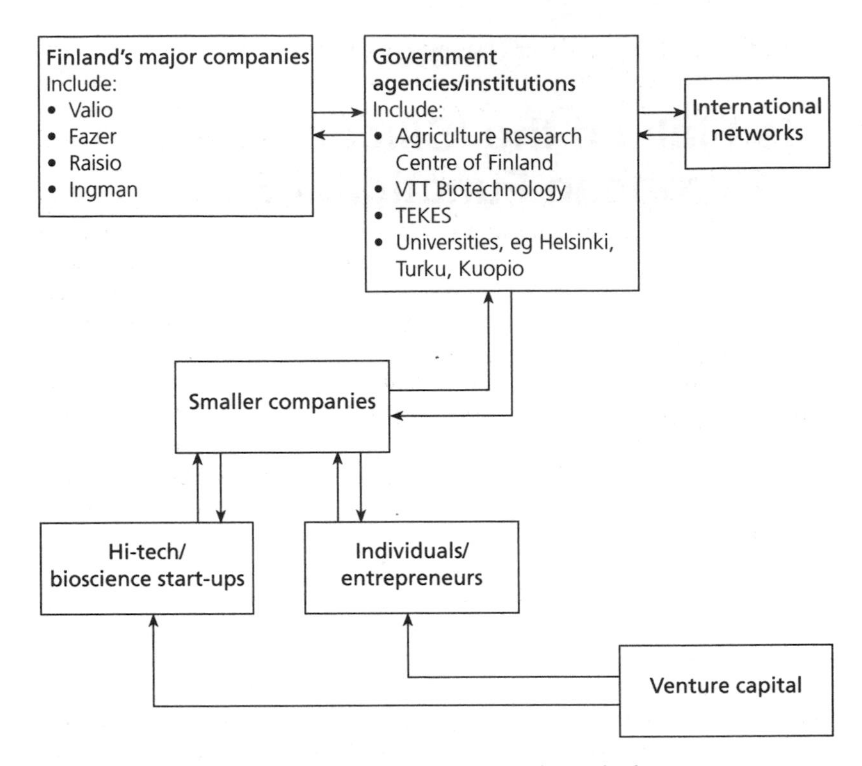

Figure 8.1 *Finland's Functional Food Cluster*

Does Finland have a functional food 'cluster'?

Finland's functional food cluster originated in 1972 with the development of the sweetener xylitol. Xylitol, a sugar alcohol derived from wood pulp, has been clinically demonstrated to be good for teeth or 'tooth friendly', helping to prevent dental caries as well as having a mouth 'cooling' effect. It is so widely used today that it is almost forgotten as one of the world's first successful functional food ingredients. The Finnish sugar company Finnsugar, which later became the Cultor Group, developed xylitol and is still the world's leading manufacturer. In 1975, the Finnish company Leaf (originally called Hellas), a part of the Huhtamaki Group, became the first company in the world to introduce xylitol chewing gum. Leaf went on to become the world's largest maker of xylitol confectionery. In Finland xylitol chewing gum has almost completely replaced traditional sugar chewing gum. But today both Cultor and Leaf are no longer Finnish-owned companies.

Other early functional innovations from Finland include the salt alternative Pansalt, invented by Professor Heikki Karppanen at the end of the 1980s and marketed by the Finnish pharmaceutical company Orion Group. Pansalt is used in more than 1000 food products in Finland and has a 40 per cent share of Finland's retail salt sales. Yet another pioneer has been Finland's largest dairy company Valio with the development of the probiotic lactic acid bacteria *Lactobacillus GG* (see below). In 1990 Valio became the first European company to launch a range of products based on a proven probiotic for gut health. The range has expanded to become one of the leading dairy brands in Finland, with some Gefilus products being market leader in certain categories. Valio has successfully using licensing arrangements to make LGG products available in 28 other countries. Together Valio and Finland's research institutes and universities have demonstrated a 'core competency' in the health benefits of probiotic lactic acid bacteria and technologies for their use in product development.

Finland is More than Benecol

Finland's functional food cluster rests on a wide-ranging scientific base. For example, Finland can boast a number of world-class researchers and scientists involved in food and health issues. In research and technology the cluster is underpinned by a strong and growing biotechnology sector, especially in medical applications, but which also has the potential for nutraceutical applications. But it is important to see developments in functional food in Finland in the context of the country's food industry and the impact of global competition on its future structure.

The Finnish Food Industry

Food production ranks fourth against all other industries as measured by the gross value of production (Finnish Food and Drink Industries' Federation publications, 1999), which was FIM 49 billion in 1997 (13 per cent of gross value of all Finnish industry). The food processing industry employs 40,000 people, over half of whom are female – the food sector in total, from farm to consumption, employs 320,000. Finland is more than 80 per cent self-sufficient in foodstuffs. Finland's major food industry product groups (measured by percentage value added) are:

- Meat processing and slaughtering (22.7 per cent);
- Dairy products (14.4 per cent);
- Beverages (13.5 per cent);
- Fruit and vegetables (5.5 per cent); and
- Animal feed (4.5 per cent).

Food processing companies in Finland number more than 600, but most are small family-owned bakeries. Table 8.1 lists Finland's top food companies defined as those having sales over FIM one billion. Finland's largest food companies, for example Cultor, Valio, Raisio Group, Fazer, Ingman Foods and Huhtamaki (Leaf), and a number of small, highly innovative companies, are all actively involved in the production of foods and ingredients with enhanced health benefits. But only 39 food and drinks companies were listed in Finland's top 500 companies in 1998.

To put the Finnish food industry in a global context, using the data from Table 8.1, but excluding both Cultor and Huhtamaki which now have Danish and Dutch owners respectively, the remaining 11 companies had combined sales of around US$5.4 billion in 1998. This is about the same as the 1999 sales of single companies such as General Mills or The Kellogg Company, and a fraction of the sales of food giants like Nestlé and Unilever who have annual sales in excess of US$50 billion. Only four Finnish companies – Rettig, Fazer, Raisio and Valio – had sales in 1998 greater than Kellogg's convenience food brand Pop Tarts, which achieved sales of about US$500 million in 1999!

Not all Good News for Finland's Food and Health Companies

Cultor Group was the largest Finnish-owned food company until, in March 1999, it announced its proposed merger with Danish company Danisco to form one of the world's largest food ingredients companies, with its headquarters in Copenhagen. The merger with Danisco followed a disastrous year for Cultor, when the company's stock market value and profits plummeted – in particular, operating profits for Cultor Food Science plunged by 73 per cent. The company was renamed Danisco at the end of 1999.

Cultor Food Science, a division of the Cultor Group, was a major supplier of xylitol and other healthy-eating ingredients, such as the fat replacer Benefat and bulking agent polydextrose. But Cultor was not the only Finnish company with food and health interests to feel the strain of global competition. Huhtamaki's Leaf confectionery business

Table 8.1 *Finland's Top Food Companies*

	1998 net sales FIM(m)	(US$m)	*Employees*	*Ranking*
Cultor (food ingredients) – now owned by Danisco	8437	(1.36bn)	7192	20
Valio (dairy products)	8056	(1.30bn)	4537	21
Huhtamki (Leaf confectionery 49 per cent – Leaf is now owned by CSM)	6387	(1.03bn)	9551	27
Raisio Group (arable food producer)	4947	(788)	2817	35
Fazer (confectionery, baking, food service)	4467	(721)	9397	38
Rettig (beverages)	3725	(601)	3512	40
HK Ruokatalo (meat processor)	2719	(439)	2989	54
Hartwall (beverages)	2648	(427)	5450	57
Atria (meat processor)	2362	(381)	2346	62
Saarionen (general food processor)	1190	(192)	1906	114
Primalco	1086	(175)	821	123
Ingman Foods (dairy products)	1085	(175)	583	124
Paulig (coffee, spices, teas.)	1085	(175)	482	125

Note: US$1= FIM6.20
Source: Finns in Business, July 1998

also found the going tough. Leaf describes health-enhancing confectionery as one of the company's 'core competencies' (it also produces traditional confectionery lines). As well as pioneering xylitol-based confectionery, Leaf developed the world's first reduced-calorie, reduced-fat chocolate count-line bar, called LO GO. However in 1999, Leaf's sales declined by three per cent and its chief executive officer Timo Peltola said: 'In confectionery, Leaf competes with international giants several times its size. Therefore, we have launched a review of Leaf's strategic alternatives.' At the end of April 1999 Huhtamaki announced it had sold Leaf to the Dutch company CSM.

In an increasingly competitive world food market, Finland could lose its world-beating position if its functional food cluster is not coordinated and capitalized on. We explain why, together with some solutions at the end of this chapter. This potential problem is compounded by Finland's rapidly disappearing food industry. Apart from Cultor and Leaf having new owners, during the late 1990s Valio, Finland's largest food company, was continuously at the centre of merger speculation. Fazer, one of Finland's oldest food companies, with 1999 sales of FIM 4.7 billion, and business in confectionery, food service and baking, announced on 8 December 1999 that its confectionery division, with a turnover of FIM 1.7 billion had merged with the Swedish confectionery company Cloetta. The new company, will be called Cloetta Fazer with headquarters in Stockholm, rather than Helsinki. The rest of the Fazer business remains in Finland.

There is therefore a real danger that the Finnish food industry could shrink to a collection of small bakers and sausage makers. In the global food economy, as epitomized by functional foods, there is a compelling case on the grounds of scale and resource allocation for the Finnish food and related industries to work more closely together to be competitive and remain innovative. There is pressure for the development of a coordinated strategy for the Finnish food industry. One option is to centre this around the informal functional food cluster.

World-renowned Food and Nutrition Policy

Finland's functional food expertise also rests on world-renowned public food and health policies to combat a range of diet-related diseases, especially heart disease. Finland has become widely quoted in food and nutrition policy circles for its ground-breaking approach to public policy on diet and health. Faced with the world's highest rates of death from coronary heart disease among its population in the 1970s a number of prevention strategies and interventions were devised. The main target risk factors were smoking, elevated serum cholesterol and elevated blood pressure. Most of the emphasis was on influencing the dietary and smoking habits of the population. A decrease in the high-saturated fat intake and a relative increase in unsaturated fat consumption were central goals of the dietary interventions. Over the years, attention has also been directed at the issues of physical activity, overweight, diabetes, alcohol consumption and psychosocial factors.

Most prominent among the public health strategies is what became known as the North Karelia Project. This began in the 1970s in response to a petition from local people for urgent help to reduce the extremely high levels of heart disease in this eastern province of Finland. Originally planned to run for five years, the project was extended and used as a pilot to contribute to national food and nutrition policy developments. The results in North Karelia have been significant, for example death rates from coronary heart disease for men fell by around two-thirds; both smoking and mean serum cholesterol also fell (Puska et al, 1995).

In Finland as a whole, coronary heart disease mortality declined by 55 per cent among men and 68 per cent among women between 1972 and 1992. About three-quarters of this decline has been explained by changes in the main coronary risk factors (smoking, blood pressure and serum cholesterol), the decrease in serum cholesterol being the most important (Pietinen et al, 1996). Pietinen et al (1996) describe the decline in coronary heart disease mortality in Finland as remarkable during the past 25 years. But the rates are still relatively high compared to other countries and prevention of the disease by dietary means is still the most important aim of food and nutrition policy in Finland. Cardiovascular disease remains the most common cause of death accounting for almost half of all deaths in Finland (National Public Health Institute, 1999).

The Finnish diet, which used to be high in dairy fat and salt, has now one of the lowest fat contents reported in Europe and has an average salt content (Pietinen et al, 1996; Valsta, 1999). In the early 1970s saturated fats accounted for 21 per cent of total energy take; by 1982 the figure had dropped to 19 per cent and by 1997 to 14 per cent (National Public Health Institute,1999).

As a result of their diet- and health-related problems, Finns have been particularly drawn to the concept of functional food as a potential public policy measure. For example, the Agricultural Research Centre of Finland has increasingly focused its studies on the development of food with health-promoting qualities. But the attention given to functional foods and their role in public health is far from being universally accepted by scientists, policy makers and, most importantly, consumers. For example, a study compiled by Finland's National Consumer Research Centre, and published in 1999, found that Finnish consumers are not only wary of functional foods and their stated claims, but on average believe that nothing substantial can be gained from their consumption, which would not be found in a healthy balanced diet. According to the report, 80 per cent of consumers surveyed believed that functional foods are unnecessary for people who exercise and enjoy a balanced, low-fat diet. Only 15 per cent of those surveyed felt functional foods were beneficial.

We have asked a number of Finnish food companies about the influence of nutrition and public health policy in relation to food and diet in their thinking and development of functional foods. To date, no one has admitted to this as a direct influence, but many concede it could have played a part in the consumer acceptance of functional foods. From our observations, we believe the link between public health and policy interventions is related to the unusual level of creative industry activity in food and health, even if it has only raised a nutritionally literate and aware consumer to stimulate food industry product development. But we have no empirical evidence to support our point of view or to evaluate the nature and importance of such a link, let alone how it might work.

Finland's public policy initiatives in food and health policy show that it is possible to combat disease and illness through dietary interventions and importantly, that it is a long-term effort for population dietary change to take place and for change to be sustainable. In the case of Finland, it has taken from 20 to 25 years to achieve its remarkable results.

Building a Strong Science and Technology Base

There are a number of dedicated research programmes supporting Finland's functional food industry cluster. Foremost is VTT Biotechnology, one of Finland's nine Technical Research Centre of Finland (VTT) institutes, funded primarily by the public sector and EU projects, but increasingly from private sector contracts. In 1999 it had an annual budget of FIM 100 million and employed about 300 people. In 1997 it started a four-year Future Foods Programme with a budget of FIM 35 million. The objective of the programme was to develop new products and processes that improve the nutritional and eating quality of foods and it now forms about 10 per cent of its annual research activities.

The focus of the Future Foods Programme is on bioprocessing, in particular lactic acid bacteria technology, enzyme technology, technology of living plant cells, novel or 'intelligent' packaging, and consumer acceptability. Professor Kaisa Poutanen, head of the programme, says that the aim is not to add new substances to nutritionally poor foods, but to use healthy raw materials, preserving and improving their positive characteristics throughout the manufacturing process. This is a somewhat unique vision for functional food in an international context.

It is worth quoting at length Professor Poutanen's vision for health-promoting foods and the thinking behind VTT's Future Foods Programme:

> *'To consumers, food represents a lot more than a source of energy: it's a source of pleasure, a means to express values in life, a means to take control over health. Safety and pleasantness can be assessed directly; many of the other food determinants must be conveyed by informa-*

tion to the consumer. This is a challenge both for the food industry and food authorities. The consumers wish for convenience and enjoyment in their eating patterns, but at the same time they want healthiness in an easy way. The food industry will provide many new alternatives for constructing a healthy diet. Yet, hedonics and ethical values will be highly appreciated.

'As we learn more about our genetic background and the risks related to certain diseases, each individual will hold more personal responsibility for his or her well-being. The food industry will offer foods targeted directly to certain risk groups. These foods will have distinct health claims with proven influences on certain biomarkers related to reduced risk levels of a particular disease. On the other hand there will be more "ordinary foods with a healthy option", for example foods based on unrefined, optimally processed plant raw materials. Ingredients will be carefully selected and processing will be tailored to maximally exploit the natural bioactivity of the raw materials. In the long run this also means that "normal" food will become more and more healthy.

'The trends mentioned above signify that functional foods are here to stay, but they will be differentiated to those specifically providing prevention for certain disorders and diseases, and to those supporting healthy choices in general. The former type of food products demand higher product development costs and will be few in number. The latter will become more abundant, but will be successful only when the sensory quality equals or is superior to the conventional alternatives.' (Poutanen, 2000)

Figure 8.2 represents how Poutanen sees the food science and technology linkages in the functional foods revolution.

VTT Biotechnology has also coordinated a four-year EU-funded project – the 'Demonstration of Nutritional Functionality of Probiotic Foods', headed by Professor Tina Mattila-Sandholm. The project involved eight EU countries and a number of industrial partners including Swedish dairy company Arla, Finnish dairy company Valio, Nestlé, and Danish company Chr. Hansen, one of the world's largest suppliers of lactic acid bacteria.

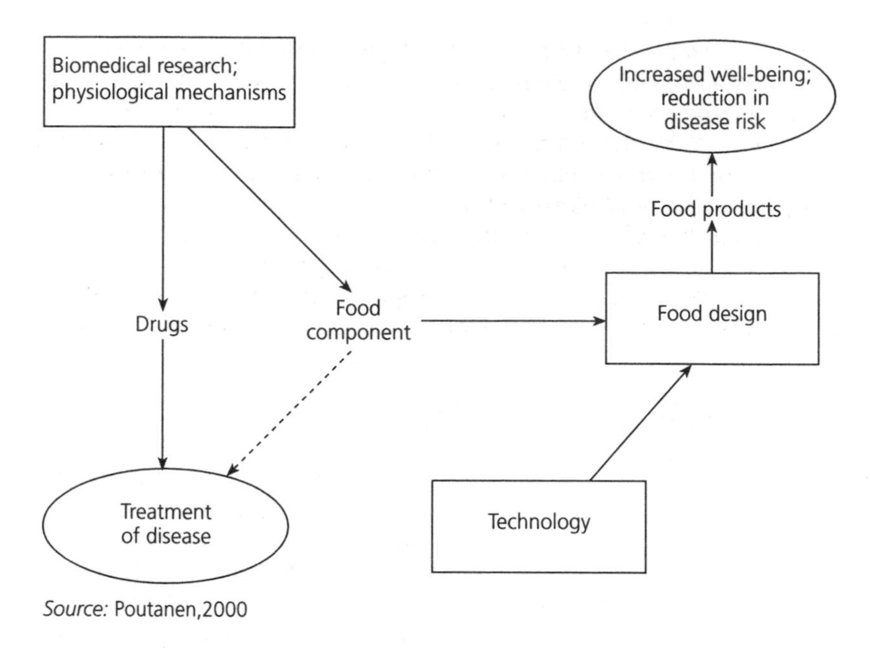

Source: Poutanen,2000

Figure 8.2 *The Research–Food Interface*

The programme investigated seven scientifically selected live lactic acid bacteria (probiotics) from their safety, through to pilot-scale food manufacture. One of the aims of the programme is to enable industry to demonstrate health claims for probiotic products in relation to the management of intestinal disorders and immune enhancement.

Rye: the Bread of Life

One of the longest-serving industry-funded nutrition information services has been Finnish Bread Information (FBI). FBI promotes bread and cereal products as part of a balanced diet. Founded in 1961 as the Bread Commission, it changed its name to FBI in 1989, and provides a wealth of information and educational materials on diet and health. For example, 'Bread Week' has been an annual event promoting bread and cereal products in Finland since 1964. Over a 10-year period (up to 1999), FBI had distributed 1.5 million of its nutrition leaflets and booklets, in a country of just five million people.

Since 1993 the FBI has been actively involved in the Nordic rye bread project. Rye is being promoted as 'a traditional concept for functional food'. The rye project has involved researchers from

Finland's universities, VTT Biotechnology and major bakery companies, to research the health benefits of rye. The project includes other Nordic countries, notably Sweden and Denmark. Its focus has been the role of the dietary fibre complex found in rye and its role in disease prevention; it has also studied lignans and other bioactive components of rye (Kujala, 1999).

The FBI has played a central role in communicating the health benefits of rye to the public. For example, between 1994 and 1998, 828 articles based on the 'rye story' appeared in Finnish print media. Work on rye is continuing, focusing on rye and cholesterol reduction.

Promoting Business Development

Part of VTT's Future Foods Programme-funding is being supported by Finland's government-funded Technology Development Centre (TEKES). Its primary objective is to promote the competitiveness of Finnish industry and the service sector by technological means. In 1997 TEKES set up an Innovation in Foods Programme, its first initiative for the food industry, with a budget of FIM 100 million. The aim of the programme is to ensure the production of increasingly high value and competitive food products. The programme is divided into two parts: new food processes, logistics, process-related measuring techniques and automation, and foods and health.

The programme claims to have 93 companies and organizations on board and another 14 participating universities and research institutes (not all the projects are directly related to functional food). It could be seen as formalizing a functional food cluster in Finland, but this is not a stated aim and it ended in 2000.

The programme has already supported two functional food companies and the entry of two products into the Finnish market. The first is Yosa, a fermented oat-based probiotic dairy-alternative snack or dessert produced by Bioferme, which was previously a small family-run business, best known for producing organic fermented products such as organic sauerkraut. Yosa was first launched in 1995 and is now available in all Finland's major supermarkets – it has achieved annual sales of more than FIM 6 million. Figure 8.3 illustrates the mini-cluster behind the successful development of Yosa.

The second company to be supported by the programme is Omecol, a company set up in 1997 to market a patented process to change the fatty acid content of foods so that they contain mostly unsaturated fatty acids. In early 1999, a version of Finland's traditional spiral sausage was the first product to be launched using the Omecol method. Called Balanssi, the product has been developed by

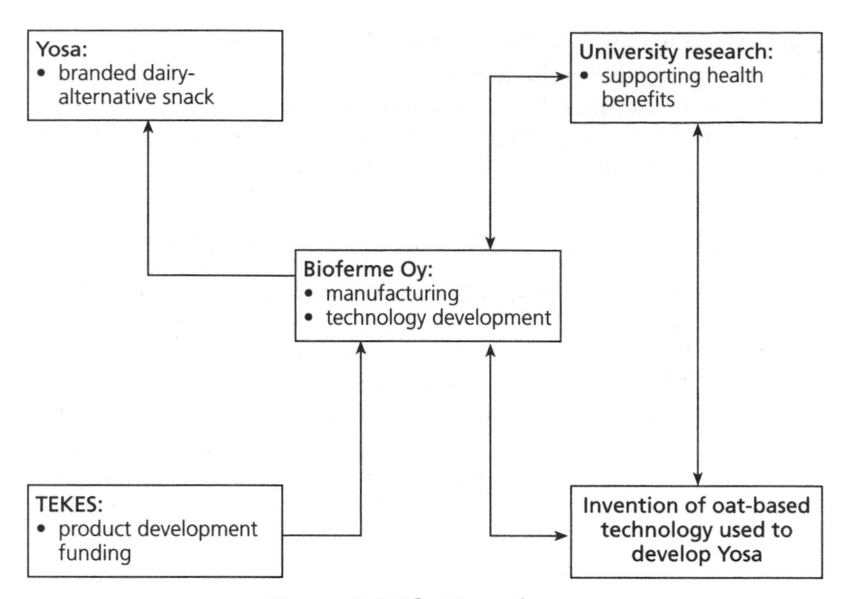

Figure 8.3 *The Yosa cluster*

Huittisten Lihapodat, Finland's largest privately owned sausage maker with roughly eight per cent share of the Finnish sausage market. The sausage packaging includes Omecol branding, logo and labelling which reads 'Food products produced by the Omecol method help to lower cholesterol together with a healthy diet'. Huittisten launched Omecol-method meat balls in May 1999.

Valio Pioneers International Probiotic Growth

Finland's two largest dairy companies, Valio and Ingman Foods, have been pioneering the commercialization of two of the world's most scientifically researched probiotic lactic acid bacteria.

In particular Valio, which accounts for aboout 70 per cent of the Finnish dairy market, has been steadily building up the international market for its probiotic *Lactobacillus Goldin and Gorbach* (LGG). Valio has been the pioneering company in functional food in Finland using probiotics over the past decade, but has not commanded the high profile Raisio achieved with Benecol. But Valio's story with LGG also illustrates that there is no instant success in this market; the company has exercised great patience to achieve market success.

Valio's work on probiotics started in the early 1980s when their researchers believed that probiotic lactic acid bacteria offered considerable potential for healthy-eating product development – a radical idea at the time. So the company approached dairy experts at Finland's universities looking for opportunities for cooperation. This route drew a blank, but Valio did learn about a human probiotic strain that had been isolated in the US, that became LGG.

A 'model' probiotic lactic acid bacteria (LAB)

LGG is a probiotic bacterial strain, *Lactobacillus rhamnosus* GG ATCC 53103, isolated from a healthy human. It was discovered by two Americans, Dr Sherwood Gorbach and Dr Barry Goldin in the early 1980s. In 1987 Valio signed a licensing agreement for the rights of LGG with the Gorbach and Goldin Corporation. LGG is Valio's trademark for *Lactobacillus GG* which is also a patented strain. Valio holds worldwide commercial rights for LGG, which has the most extensive scientific dossier of all probiotics in the world and has become what we describe as a model probiotic LAB (see Chapter 9). LGG has been shown to have the ability to:

- Remain alive and active both in food and capsules;
- Tolerate conditions in the digestive tract;
- Attach to the mucous membrane of the intestine and to bowel mucus, and to boost immune response;
- Balance the microflora of the digestive tract;
- Influence bowel metabolism and to impede the formation of harmful compounds;
- To prevent and treat different types of diarrhoea;
- Repair mucous membrane damage and the immune barrier of the mucous membrane; and
- Accelerate recovery from milk allergy and reduce atopic symptoms.

The world's first LGG products were the branded Gefilus range of dairy products and juices, first introduced in Finland in 1990. Since then, Valio have licensed LGG for use by companies in 27 countries (up to May 2000) (see Table 8.2).

A Vision of Global LGG

The commercialisation of LGG, however, has not been straightforward and hides a considerable amount of faith and belief within Valio to keep the project alive. As Jaakko Lehtonen, at the time vice presi-

Table 8.2 *Products Containing LGG*

Country	*Brand*	*Products*
Europe		
Austria	1x taglich	Dairy products
Bosnia-Herzegovina	Dukat BioAktiv	Dairy products
Croatia	Dukat BioAktiv	Dairy products
Estonia	Gefilus	Dairy products Capsules
Finland	Gefilus	Dairy products Juices Capsules
Germany	Emmfit	Dairy products
Iceland	LGG+	Dairy products
Italy	Dicoflor, Floridral, Giflorex	Powdered products
	Vivi Vivo	Dairy products
The Netherlands	Mona Vifit	Dairy products
Norway	Tine Biola	Dairy products
Slovenia	Dukat Aktifit Plus	Dairy products
Switzerland	Emmi Aktifit Plus	Dairy products
UK	Campina Vifit	Dairy products
	Culturelle	Capsules
Middle East		
Israel	Tnuva LGG1	Dairy products
Asia		
Japan	Onaka-He-GG	Dairy products
South Korea	Maeil GG	Dairy products
Oceania		
Australia	Vaalia	Dairy products
South America		
Argentina	La Serenisima Serenito Ser	Dairy products
Brazil	Batavo GG	Dairy products
Boliva	Next	Dairy products
Chile	Next Uno al Dia	Dairy products
Ecuador	Toni	Dairy products
North America		
US	Culturelle	Capsules

Source: Valio, 2000

dent of Valio Fresh Products and responsible for the world-wide commercialization of LGG told us in an interview in 1998: 'We have a vision of global LGG, but five years ago that sounded stupid – we did not have one contract in place.' He explained:

> *'In the early 1980s people had a vision and became committed to probiotics as an industry for the future. These people have now been shown to be correct, but it took 10 years before we got the first results. You just had to believe in it. Other companies have now got into this area, in particular Nestlé, Danone and Yakult. They helped us a lot because they developed the market. Now there is a market and since research and development takes years and years, other companies are opting for licensing and started to approach us – with a licensing agreement they also buy time.'*

In August 1999, Valio joined the 'battle of the little bottles', launching its own 'daily dose' format in Finland under its Gefilus brand (see Chapter 9). However, LGG licensees have also been successful in developing the 'little bottle' concept. For example in Switzerland, dairy company Emmi Frishprodukte introduced a 65 ml drink in 1996 and in 1998, Icelandic company Mjolkursamsalan also launched a LGG mini-drink – selling six million bottles in its first year to Iceland's population of just 270,000.

Dairy Company Aims to Lead Finland with Functional Foods 'Image'

Finland's second largest dairy company, Ingman Foods, has followed Valio into the functional food market, but through working with a specialist Swedish research company BioGaia Biologics and developing a range of products based on licensing the use of BioGaia's probiotic *L. reuteri*. Ingman's marketing director, Marika Ingman, whose grandfather founded the company at the turn of the 20th century, is banking on functional foods to spearhead the company in the 21st century. In 1997 the company launched a new brand RELA (in Finnish 'rela' is a colloquial expression for 'relax') to develop a range of products which Finnish consumers will come to recognize for their health benefits.

While the RELA brand is a growing market niche, Marika Ingman also saw the brand as reflecting the type of image the company as a whole wants to project for the future. At the time the product range consisted of four products – fat-free yogurt, fermented milk, pineapple and orange juice, and cottage cheese – but Ingman are investing a lot in the brand and plan to develop 20–30 RELA products with health benefits.

> *'But first we have to convince and educate the consumer. People over 30 are eating less and thinking about the quality of what they eat, not only price. As people think more about what they are eating they want products which are really good for them,'*

said Marika Ingman. She is also keen to ensure that the new company gets its message about functional foods right and is very careful about what claims it makes. Currently Ingman Foods is building the RELA brand around the basic message that 'RELA helps you to stand stress', as advertised for the first time on television in 1998.

Marika Ingman sees functional food as a long-term investment, saying:

> *'We need companies to come into the functional foods business to be serious. Even if regulation is strict, the best companies will go that bit further because they have invested a lot in research. With functional foods we need to proceed slowly. With functional foods it is a step-by-step process, a long-term commitment.'*

Two years after launch, the RELA brand is catching up with Valio's Gefilus probiotic yogurt. Ingman are also hitting their target market of mainly women aged 25 to 45 and living in Finland's bigger cities.

Marika Ingman says the company has ambitious plans for their functional foods brand – 'We want to be "image leader" for functional foods in Finland.'

Continuing Finnish Functional Foods Creativity

Finnish companies continue to be innovative in the functional food market. In April 1999 Finland's oldest and second largest soft drinks company Sinebrychoff launched Finland's first probiotic soft drink. Called Hyvaa Paivaa (Good Day), it is a fruit beverage containing a mix of nine different fruits, fibre, the antioxidant vitamins A, C and E and added calcium (one bottle contains 31 per cent of the recommended daily intake of calcium). But its key health message is that the beverage is a prebiotic. On packaging and supporting point-of-sale literature the company actually talks about prebiotics, describing them as 'nature's nutrition fibres that improve your gut and stomach function'.

The fibre in Good Day consists of oats, pectin and inulin. Product packaging recommends a dose per adult of one bottle a day, which delivers five grams of fibre (and also contains a warning that too much could have a laxative effect). Product literature goes on to say that prebiotic fibres:

> *'strengthen and nourish the good bacteria in your gut and function as the gut's "plumber" by transporting foreign substances through the gastrointestinal system. Your stomach will adapt to the "awakening" of the beneficial bacteria very quickly'.*

Apparently Good Day has become a big hit among Finnish lorry drivers!

The Future: Multi-Bene

As for the next Finnish functional food ingredient 'blockbuster', we believe Professor Heikki Karppanen at the University of Helsinki, already famous as the inventor of the successful salt alternative Pan Salt, has the answer. His latest ingredient invention is Multi-Bene, a patented food composition, with the international patent application stating 52 claims.

Put simply, Multi-Bene is a combination of natural plant sterols with the minerals calcium, magnesium and potassium at relatively low levels. As an ingredient, Multi-Bene can be incorporated in many common food items such as bread, hamburgers, pizzas, yogurt and meat products.

The potential health benefits of Multi-Bene include an effective lowering of elevated serum cholesterol, lowering of blood pressure, decrease in obesity, and improvement of glucose metabolism/insulin sensitivity. These are all areas that define the 'metabolic syndrome', also known as Syndrome X in the US. Metabolic syndrome is a very common problem, especially among men with a lot of fat around their waists – 20 to 30 per cent of Finnish men, for example, suffer from it. If Multi-Bene is scientifically validated as treating or preventing metabolic syndrome it would be a major breakthrough and be of benefit to literally millions of adults. Currently a number of human clinical trials are being carried out and the results will be published in peer-reviewed publications.

Food products containing Multi-Bene are planned for launch in Scandinavian countries in 2001. Finnish food company Fazer, now Scandinavia's largest confectionery company, Pouttu, a specialist meat company, and Valio have all signed deals in their respective product areas to develop products using Multi-Bene for markets in Finland, Sweden, Norway, Denmark and Russia. However, on the recommendation of Finland's food authority, all three companies have agreed for Multi-Bene to go through the EU novel foods process, and will await the outcome before introducing products.

The Future of Functional Food in Finland

In this chapter we have written glowingly about functional food developments in Finland based on the many exciting product and ingredient activities that are now taking place. The level of activity, and the range of skills and technical competencies, suggest that Finland has a true geographic concentration of expertise that has been built up over the long term. It would be easy to assume that Finland has a successful functional food cluster, and even that it is the 'Silicon Valley' of functional foods. But a prerequisite for cluster status is the presence of interconnected or linked industries and other entities important to competition (Porter 1998a, 1998b). Despite Finnish food industry success in some areas this prerequisite is poorly developed, and results in an underdeveloped and underperforming cluster. The cluster is weak for six reasons:

1 A lack of capital;
2 Poor business development and marketing skills;
3 A lack of leadership;
4 Poor sharing of knowledge and specialized information;
5 Lack of cooperation;
6 Difficulties in achieving international scale.

The strengths of the cluster, however, are:

1 The presence of scientific know-how and research expertise;
2 Successful functional food product development experience;
3 Already involved in world-wide marketing of functional foods and ingredients;
4 A long history of functional food and ingredient involvement to draw upon;

5 Highly inventive and innovative individuals in science and technologies;
6 A growing biotechnology cluster to support the functional foods and nutraceuticals cluster.

As Porter points out, however, all industries can employ advanced technology, and all industries can be knowledge-intensive. What is needed to gain the advantages of a local cluster is the quality of the business environment. In this respect the weaknesses identified above could be crucial.

For example, one of the biggest surprises from our research was the level of secrecy and lack of cooperation in Finland's food industry. At first glance it seems the opposite is true. As a small country Finland has well-established professional networks and key people who all know each other. There are instances where, through word of mouth, companies have developed new products – but there it ends. We have been told on many occasions that it is nearly impossible to get the major companies to talk with each other, even in areas of mutual interest, and many innovators are reluctant to seek help in developing their idea through fear of it being 'stolen'. If this seems fanciful, we were told, in all seriousness, by one Finnish venture capitalist that his company has been approached by people seeking funds for functional food and nutraceutical projects, but who will not reveal the barest detail of their invention or idea.

A key advantage of a 'cluster' is access to specialized information (Porter 1998a). As extensive market, technical, and competitive information accumulates within a cluster, members have preferred access to it. In addition, Porter says, personal relationships and community ties foster trust and facilitate flow of information. This again appears to be poorly developed in Finland's food industry.

Porter's model also highlights how one or two companies can assume a leadership role that stimulates innovation and growth for many others. This is not happening in Finland's functional food cluster, there is no functional food 'Nokia' in Finland. In fact, the opposite may be starting to occur because as the potential for functional food unfolds, more and more people want to take a slice of the collective cake for themselves, diluting and weakening resources and collective action.

Finns Need to Get 'Talking'

While there is financial and institutional support for science and technology for Finland's functional food cluster, a weakness identified by a number of people we have interviewed is poor business development and marketing skills to pursue international market opportunities. Another factor is limited access to capital to fully exploit functional food science and technologies internationally. Finnish companies have largely relied on licensing deals for international business development.

This in practice is seeing Finnish companies making the investments in terms of R&D and technology resources and risk – in other words the basic science and raw material supply – but missing out investing and capitalizing on these in consumer markets, other than within Finland. This is enabling others to reap the full benefits of functional food success in consumer markets. For example, within European probiotics, as Chapter 9 explains, Yakult, Nestlé and Danone are driving consumer markets, whereas Valio, the European pioneer in this area, is an unknown name except within Finland.

A robust functional food cluster would suggest Finnish companies and entrepreneurs might be gaining more financially from their internationally recognized intellectual capital, for example with more business start-ups, and international presence, in consumer markets. The licensing route, while seductive, is fraught with difficulties as the Raisio and Benecol story shows (see chapter 2). Finally, a healthy functional food cluster should more readily attract capital – a key characteristic of the real Silicon Valley is the operation of its venture capitalists.

Finland may still be able to become the 'Silicon Valley' of functional food' within Europe (although the Swedes are fast developing their own functional food 'cluster' that will soon eclipse Finland on present trends). However, Finland has the building blocks already in place and clearly has extensive functional food expertise, technologies, and an international edge; the weaknesses to be addressed are far from insurmountable. Porter's work suggests a way forward – government policy and industrial policy should look to build on existing and emerging clusters. There would therefore have to be a new agenda of collective action in the private sector for those businesses and companies involved in functional food. Cluster case-studies indicate they require a decade or more to develop depth and competitive advantage. A real strength of the Finns, however, is their ability to take a long-term view – a prerequisite for functional food success, but the Finnish food industry and others involved in the area will have to get 'talking' – something Finns are not always renowned for.

Photo 1 *Pro Viva, the first-ever probiotic juice drink, was a joint development between a science-based company and a dairy. It became a hit product in Sweden and can now be found in 14 European countries*

Company publicity photograph

Photo 2 *Benecol, the first-ever cholesterol-lowering margarine. The company behind it, the Finnish Raisio Group, saw its shares become the subject of a stock-market frenzy*

Company publicity photograph

Company publicity photograph

Photo 3 *Finland's Valio has successfully commercialized the probiotic LGG in dairy products around the world. LGG can be found in Finland in Valio's successful Gefilus range*

Photo 4 *The 'FOSHU' symbol, carried on foods approved under Japan's regulatory system for functional foods, the first such system in the world*

Company publicity photograph

Company publicity photograph

Photo 5 *Yakult sells 26 million bottles of its fermented milk drink every day in 18 countries around the world. The company entered the US market in 1999*

Company publicity photograph

Photo 6 *Novartis' Aviva range – a complete range of functional foods launched in 1999*

Company publicity photograph

Photo 7 *Tropicana's calcium-fortified juice became a hit product in the US*

Company publicity photograph

Photo 8 *Danone's Actimel, a fermented milk drink on sale in Europe and the US*

Company publicity photograph

Photo 9 *Flora Pro.Activ – Unilever's cholesterol-lowering margarine, launched in response to Benecol*

Company publicity photograph

Photo 10 *Novartis have adapted the 'little bottle' concept and introduced a fruit juice-based morning drink called Oclea, which has proved successful in Switzerland*

Company publicity photograph

Photo 11 *The first food-specific health claim approved by the US Food and Drug Administration was for soluble fibre and heart health following a petition from the Quaker Oats Company*

Part 4

Global Marketing and Strategic Challenges

Chapter 9

The Marketing of Functional Food and the European Dairy Industry

Imagine yourself in charge of a company, which is about to enter the European market for the first time. Your company and brand are unknown. You sell one product, in one flavour, in one size, in one packaging format. You do not offer a private label, or discounts or price promotions. Your product is completely unique in Europe and an entirely new and unknown concept to consumers. You keep advertising to a minimum and prefer to communicate to consumers face-to-face. And by the way, the selling proposition is that your product is good for the digestive system.

It sounds impossible – and yet this was precisely the position which faced Japan's Yakult when, back in 1994, it made its entry into the European market with its 65 ml 'daily dose' fermented skimmed milk drink, based on the probiotic lactic acid bacteria *Lactobacillus casei Shirota*. It is hardly surprising that at the time many people in the European dairy industry expressed the view that Yakult would not succeed.

Five years later and Yakult is selling around 550,000 bottles a day in The Netherlands, Belgium, UK and Germany – making it a US$80 million plus brand in Europe in 1999. Yakult have plans to introduce Yakult to other European countries. And, if imitation is the sincerest form of flattery, then Yakult can feel flattered indeed, for its success has spawned an ever-expanding following of 'lookalikes' across Europe and the world.

Globalization of Food and Health

Developments in the European dairy industry are revealing because they characterize the trends leading to the rapid globalization of food and health, which we set out in Chapter 1, and they must be viewed in the broader context of change within the food industry and the exploitation of food and health trends. To recap, these globalizing trends are:

- The rapid diffusion of functional food technology globally;
- The closing international gap between basic science and rapid commercial development;
- The range of challenges facing companies working in this area are also being met internationally, from regulatory hurdles to technology and business development.

As our description of the business activities within the dairy sector in Europe will show, there is a ferment of activity, not just to secure local markets, but to create or capture international growth and value through functional food. It is a sign of the speed and power of these very real forces that we open this chapter on European dairy with a description of the market-transforming activities of a Japanese company. But we are jumping ahead.

Structure of this Chapter

This chapter is structured in the following way. First, the science behind probiotic lactic acid bacteria and the concepts of prebiotics and synbiotics are briefly introduced and defined to explain the scientific background to functional food marketing in the dairy industry. Second, the significant areas of marketing activity in recent years in products using probiotics, prebiotics and synbiotics is analysed using three case studies:

1. The 'battle of the little bottles' – one of the highest profile areas of functional food marketing in Europe has been the introduction and success of tiny (usually 65 ml) 'daily dose' bottles of fermented milk drinks containing specific strains of lactic acid bacteria with documented health effects;
2. Small is beautiful for Sweden – we profile the activities of two research-based companies set up in the 1990s to develop and promote a particular strain of probiotic lactic acid bacteria. We detail how one Swedish dairy company is using one of these company's expertise to create a value-added market for a probiotic juice drink;
3. Developments in European probiotic yogurts with health benefits – we examine the marketing of a range of functional yogurts in The Netherlands and France as illustrative of general trends in Europe for yogurts based on probiotic lactic acid bacteria and the concept of prebiotics and synbiotics. We also look at the case of the Danish company, MD Foods, and how it created a 'hit' yogurt brand with a proven cholesterol-lowering claim in Denmark, but saw the brand's claim meet regulatory trouble in the UK.

Finally, we conclude with the lessons from the European experience for the US, not only because the US dairy industry is on the verge of transformation similar to that which Europe has undergone (a transformation likely to be driven in America by French, Danish, Swedish and Japanese companies) but because the lessons of the European experience are applicable across many markets.

The Scientific Background to Functional Food Dairy Industry Marketing Developments in Europe

The definitions of 'probiotics', 'prebiotics' and 'synbiotics'

For thousands of years microbial cultures have been used to ferment foods and prepare alcoholic beverages. Food products such as soy sauce and pickles, salami-type sausages and dairy products such as yogurt and some cheeses, as well as alcohol production have relied upon the fermentation properties of lactic acid bacteria (LAB) for their production. But it was only in 1857, while investigating an industrial problem besetting the alcohol producers of Lille in France, who produced alcohol made from sugar beet, that Louis Pasteur, professor of chemistry at the University of Lille, first identified the role of lactic acid bacteria in fermentation. It was not until 1965 that the term 'probiotic' was first used by Lilley and Stillwell (1965) to describe beneficial micro-organisms.

Within Europe, one of the main areas of scientific research over the past decade in functional food science has been a renewed focus on the role of the gut in human health. In particular, research has focused on the billions of live bacteria that colonize the human gut and the positive and negative roles these might have on human health. It is truly a new and evolving area of scientific investigation. In fact a recent review on the topic said:

> *'One of the most promising areas for the development of functional foods lies in modification of the activity of the gastrointestinal tract by use of probiotics, prebiotics and synbiotics.'* (Salminen et al, 1998a)

Within European markets, but especially in Northern European countries such as Sweden, Finland, The Netherlands, France, Germany, Belgium, Denmark and the UK, the health of the gut has already become one of the major functional food markets based on products using probiotics, prebiotics and synbiotics as ingredients.

What are probiotics?

When we are first born, the gut contains no microflora (because the womb is a sterile environment) and the large-gut microflora is only acquired after birth. The species of gut bacteria that develop are largely controlled by the type of diet, since it has been shown that the composition of the gut microflora will vary depending if the baby is breast-fed or bottle-fed. After weaning, a pattern that resembles the adult microflora becomes established.

In adults, the gastrointestinal microflora represents an ecosystem of the highest complexity (see Holzapfel, et al, 1998). The gastrointestinal tract of an adult human is estimated to harbour 10^{14} viable bacteria; this equates to about 95 per cent of all cells in the body. It is estimated that the colon of healthy adults contains about 300 to 400 different cultivable species belonging to more than 190 genera. In general terms, intestinal bacteria may be divided into species that are either harmful or beneficial towards host welfare. Through the process of fermentation, different colonic bacteria are able to produce a wide range of compounds that have positive and negative effects on gut physiology.

The scientific (and commercial) goal has been to identify and use, or increase the number of, positive or health-promoting lactic acid bacteria, often at the expense of those bacteria producing a negative effect. Spearheading dairy functional foods have been products that help the health-promoting bacteria that live in the human gut.[1]

'Probiotics defined

Most prominent in terms of the marketing and product development of functional food in the dairy industry has been the use of probiotics. The classic definition of probiotics is 'live microbial feed supplements which beneficially affect the host animal by improving its intestinal microbial balance'(Fuller, 1991). Evidence is accumulating from well-designed, randomized and placebo-controlled double-blind studies that a few well characterized LAB strains have documented probiotic health-promoting effects when defined doses are administered (Salminen et al, 1998b).

The concept behind increased interest in products containing viable probiotic LAB is the possibility of manipulating the composition of the gut microflora. The aim is to increase the numbers and activities of those micro-organisms suggested to possess health-promoting properties such as *Bifidobacterium* and *Lactobacillus* species.

Some of the areas where probiotic LAB have been investigated and may have a health role (Goldin, 1998) are shown in the following list, although it should be noted most studies have been conducted on adults and children with intestinal or other disorders.

1 Intestinal disorders
  Diarrhoea; antibiotic induced; travellers; infantile; constipation; colitis; salmonella and shingella infections; lactose intolerance; flatulence
2 Other disorders
  Vaginitis; alcohol-induced liver disease; cancer; hypercholesteraemia
3 Other uses
  Stabilization of gut flora; recolonization of bowel after antibiotic treatment; treatment for food allergies; adjuvant for vaccines

It is not our purpose here to try to review or comment on the science behind probiotic LAB, other than to highlight the increasing knowledge underlying the importance of the role of intestinal flora in maintaining health and in disease prevention. For example, as Holzapfel et al (1998) conclude:

> *'Probiotics offer dietary means to support the balance of intestinal flora. They may be used to counteract local immunological dysfunctions, to stabilize the gut mucusal barrier function, to prevent infectious succession of pathogenic micro-organisms or to influence intestinal metabolism.'*

The authors point out, however, that many of the proposed mechanisms have still to be validated in controlled clinical trials.

In terms of the marketing of probiotic products it is important to distinguish between probiotic LAB that have been shown to reach and colonize the human gut and have health-promoting benefits and many 'live' or 'bio' yogurts that are on sale in countries such as the UK for example, where the LAB is used as a food processing aid. In these cases there is usually limited evidence that the LAB are alive in the product as consumed and/or in sufficient numbers to reach the lower gut, and that they have any proven human health benefits. For example, in a paper examining the therapeutic potential and survival in yogurt of certain LAB, Kailasapathy and Rybka (1997) cite research from Australia. They say:

> *'The range of yogurts and dried yogurt preparations available in Australia that contain probiotics is large.*

> *However, evaluation by the Dairy Research Laboratory showed that probiotic organisms are often not at high levels. Strains of bifidobacteria used in some commercial products neither survive product acidity during storage nor gastric transit. It is considered misleading to describe probiotic yogurt as having health promoting properties unless the minimum level of viable cells is present at the expiry date.'*

In other words, not all probiotics are equal in their health effects and research will differentiate them. A number of scientifically validated probiotic LAB have been identified and are being used to develop functional foods. In terms of food products that aim to promote a scientifically validated health benefit, a key requirement for probiotic LAB should be their ability to reach the lower gut and to temporarily colonize it. This in turn, depends on what are called the 'colony forming units' (CFU) available, that is, the number of health-promoting LAB in the product. Although variable for particular LAB strains, a figure of 10^6 to 10^9 CFU seems to be a 'gold standard' for product developers to aim for.

The Concepts of 'Prebiotics' and 'Synbiotics'

The concept of 'prebiotics' is an even more modern invention than 'probiotics', being first introduced in a paper by Gibson and Roberfroid as recently as 1995. In this paper they define a prebiotic as:

> *'a non-digestible food ingredient that beneficially affects the host by selectively stimulating the growth and/or activity of one or a limited number of bacteria in the colon, and thus improves host health'.*

By a 'limited number' they mean beneficial bacteria.

While many substances have been suggested as prebiotics, the most convincing science to date are prebiotic ingredients called fructo-oligosaccharides. The most popular, and those that fulfill the criteria of prebiotics in current food use, are the chicory fructans (inulin and its enzymatic hydrolysate, oligofructose), although many other ingredients are under investigation to demonstrate a scientifically validated prebiotic effect. An issue not addressed here is that prebiotics, such as the range of oligosaccharides, are increasingly regarded as having the properties of other dietary fibres, and their subsequent role in human health in this respect is also being investigated.

Orafti, a Belgium-based pioneering company in the development and scientific validation of prebiotics derived from chicory, has supported an extensive research programme into the health benefits of inulin and oligofructose. In Europe, more than 700 products now contain Orafti's ingredients, but not all necessarily for their prebiotic health benefits – inulin, for example, can also be used to replace fat in products.

In the same 1995 paper Gibson and Roberfroid also introduced the concept of 'synbiotics', defining them as:

> *'a mixture of probiotics and prebiotics that beneficially affects the host by improving the survival and implantation of live microbial dietary supplements in the gastrointestinal tract, by selectively stimulating the growth and/or by activating the metabolism of one or a limited number of health-promoting bacteria, and thus improving host welfare'.*

Table 9.1 illustrates examples of the wide range of probiotic, prebiotic and synbiotic products that have been introduced in Europe.

Examples of on-pack health claims for European functional dairy products, adapted from Coussement (1997), are given in the following list:

1 Actimel – 'reinforces your natural resistance' and 'your daily dose of natural protection'
2 BIO Aloe Vera – 'feeds and hydrates in a very self-evident way; from inside out'
3 Biotic Plus Oligofructose – 'promotes the natural balance of the gut microflora and thus your health' and 'oligofructose stimulates the body's own positive bacteria and increases (as dietary fibre) the activity and purification of the gut'
4 Fyos – 'promotes a healthy gut balance' and 'take care of your whole health'
5 Fysiq – 'contributes to a healthy cholesterol level' and 'this effect is strengthened by the presence of a dietary fibre'
6 ProCult 3 – 'a positive influence on the gut flora'
7 Silhouette Plus – 'the soluble bifiogenic fibres help preserve and re-establish the balance of the digestive flora.

Table 9.1 ***Examples of European Dairy-Based Products with a Health Claim***

Company	*Product*	*Countries*	*Type*
Aldi	Bl'AC	Germany	Synbiotic
Aldi	Biotic Plus Oligofructose	Germany/Holland	Synbiotic
Besnier	BA	Many EU countries	Probiotic
Besnier	Jour aprés Jour	France	Prebiotic
Candia	Silhouette Plus	France	Prebiotic
Danone	Actimel Casei	Many EU countries	Probiotic
Danone	Actimel Cholesterol Control	Belgium	Synbiotic
Danone	BIO	Many EU countries	Probiotic
Ehrman	Daily Fit	Germany	Synbiotic
Emmi	Actifit-Plus	Switzerland	Synbiotic
Migros	ProbioPlus	Switzerland	Synbiotic
MKW	Tuffi Vita	Germany	Probiotic
Mona	Fysiq	Holland	Synbiotic
Muller	ProCult3	Germany	Synbiotic
Nestlé	LC1	Many EU countries	Probiotic
Nutricia	Fyos	Belgium	Prebiotic
Sudmilch/ Stassano	Vifit	Many EU countries	Synbiotic
Tonilait	Symbalance	Switzerland	Synbiotic
Yakult	Yakult	Many EU countries	Probiotic

Source: Coussement (1997)

While Table 9.1 shows that many products on the European market are described as 'synbiotic' in that they contain both probiotic and prebiotic ingredients, in the strict terms of Gibson and Roberfroid's definition, this is an erroneous (if still convenient) description since the concept of 'synbiotic' still remains to be scientifically validated. By this is meant while the probiotic alone may be demonstrated as beneficial to the host, and the same for the prebiotic alone, there is as yet little scientific evidence that the two work together in 'synergy' as suggested by the synbiotic concept. This of course does not negate the individual role of the probiotic or prebiotic in the product. As Roberfroid (1998) later writes:

> *'...the concept of synbiotic has, up until now, not really been applied to the development of new foods especially functional foods. Furthermore, even when new products indeed contain a mixture of pro- and prebiotics they are seldom presented or defined as synbiotics. This concept thus remains open for validation and further research is needed.'*

The Business Background for Developments in European Dairy Markets

From a commodity to a marketing-led dairy industry?

Like the US, the European market for liquid milk and dairy products has, historically, been characterized by commodity marketing. In particular, the EU dairy market has been distorted and complicated through the price and market support mechanisms of the Common Agricultural Policy (CAP). The CAP, however, has proved so costly that there has been increasing pressure for reform and, consequently, the prospect of lower dairy prices.

Competition in the European dairy industry intensified during the 1990s with margins on liquid milk under pressure and excess capacity in the industry both at the producer level and the processor level – it is set to intensify even further.

The pressure is on Europe's dairy companies to redefine themselves in preparation for this harsher future. Companies are looking to mergers and acquisitions, to add volume to their processing, and so reduce unit costs, while also enabling them to reduce overheads. The effects can also be dramatic in market terms – the merger of The Netherlands' Friesland and Coberco dairy cooperatives in late 1996, for example, turned this combine into Europe's third largest (and the world's fifth largest) dairy company overnight. Although 15 of the world's top 23 dairy groups are European, in a future becoming even less secure, companies continue to seek ways to rationalize, expand internationally and to look to new products and markets to add value.

European dairies turn to marketing

Realizing that there will be only a few, highly visible, dairy companies left in a few years, the response of many companies has been to turn to marketing to drive growth by:

- Building portfolios of branded products and reducing their dependence on their historic generic commodity business;
- Maximizing the prices they can earn for their branded products as a way of increasing margins;
- Internationalizing their businesses, in particular taking branded products to markets they have not traditionally served; and
- Turning to innovation in ingredients to develop a presence in healthy- eating markets.

Although marketing is becoming more important for many companies, it should be emphasized that much of the dairy industry is still, often through necessity, embroiled in the commodity-style culture of old. But the functional foods revolution is leading many companies to look again at innovation. In this respect and in terms of adding value, the health of the human gut has been chosen as one of the new battlegrounds of Europe's dairy companies. All four elements listed above can be seen at work in current dairy functional food marketing strategy.

The healthy-eating opportunity

Historically dairy products have been well established, particularly in the northern European diet, and are generally regarded as 'healthy' products by consumers. But in the 1980s dairy products came under attack from nutritionists and health professionals because of their contribution to total fat intakes, especially saturated fat. Although individual dairy products are not as high in fat as some other processed foods, overall, dairy products are a significant part of the northern European diet and continue to make a significant contribution of to the total level of dietary fat.

Dairy companies vigorously defended themselves against such criticisms, but in the end many were forced to reposition themselves on healthy-eating platforms. For example, skimmed and semi-skimmed milk now dominate the liquid milk market in some countries and reduced-fat and low-fat yogurts have taken significant shares of the yogurt market in certain countries.

In the 1990s, European dairy companies started to use the increasing body of functional food science to pioneer new healthy-eating strategies. This has given dairy companies the opportunity to further position their products as an intrinsically healthy part of the diet, this time not by taking fat out, but by adding ingredients that are claimed to offer specific health benefits. The 1990s saw a ferment of market activity and by 1999 the probiotic yoghurt market was valued at almost US$1 billion, according to market research company A C Nielsen, up from US$685 million in 1997.

The importance of functional food to the dairy industry in general is neatly captured by Matti Kavetvuo, at the time chief executive officer of Finland's largest dairy company Valio. Writing in Valio's 1997 company report, he explained:

> *'The dairy sector is undergoing a rapid process of change worldwide. The conventional mode of operation, which relied on a comprehensive product range, is gradually yielding to an international and highly differentiated approach based on excellence in core areas. The leading companies in the sector are increasingly shifting their emphasis from raw materials and local markets to products that add value in the form of know-how and thus earn better return than traditional, more primary products. Functional foods are a typical example of this development.'*

The European Dairy Industry Pioneers Positive Functional Food with Products for Healthy People

An important distinction in functional foods, and one that is often overlooked, is that they are being targeted at two populations. First, consumers who have a recognized medical condition, disease or illness, such as those with elevated cholesterol levels, and second those who are healthy. With a few exceptions, the dairy industry markets functional products with positive health benefits for healthy people.

In a commercial setting a number of probiotic LAB, supported by strong scientific dossiers, are currently driving this product development – what we call the 'power probiotic LAB'. Those being successfully marketed internationally include:

- *Lactobacillus casei Shirota* from the Yakult Honsha Company and marketed in Europe from 1994;
- Nestlé's LC1 (*Lactobacillus johnsonii La1*) already available in a number of European countries, but most successfully to date in Germany;
- Danone with their Actimel products using *Lactobacillus casei Imunitass*;
- *Lactobacillus GG* from Finland's Valio and now licensed for use in 28 countries;
- *Lactobacillus plantarum 299v* from Swedish food research company Probi AB, and being successfully introduced in products by Sweden's Skanemejerier; and

- Swedish science-based company BioGaia Biologics and their LAB *Lactobacillus reuturi*.

Case Study 1: The Battle of the Little Bottles

The most prominent development in the marketing of functional dairy products in Europe has been in the area of fermented milk drinks containing probiotic LAB and sold in tiny, daily-dose bottles. However, it was not a European dairy company that created or pioneered this market, but the Japanese company Yakult Honsha (see Chapter 6). Yakult Honsha is not, as often thought, a dairy company, but has been built entirely around the science and applications of the probiotic LAB *Lactobacillus casei Shirota*.

To briefly recap, first marketed in Japan in 1955, Yakult now sells 26 million bottles a day around the world in 17 countries, making it the world's largest functional brand. There are two keys to Yakult's success. One is that the company is serious about probiotics. While some dairy companies seem to see the probiotic sector as primarily a marketing opportunity, for Yakult it is their whole business. The other is that Yakult takes a holistic approach to health, it's guiding philosophy being, 'working on a healthy society' – a point which we will expand further in Chapter 11. This approach is underpinned by a long-term perspective on their business and, most importantly, a possibly unique approach to communicating a strategy to consumers, as explained in Chapter 10.

Since 1935, when Dr Shirota first isolated the *L. casei* strain, which bears his name, the company has invested a great deal in understanding the science behind the bacteria. Interestingly, Yakult is the only company in probiotics which publicly guarantees the level of bacteria in its product – that each 65 ml bottle will contain 6.5 billion bacteria at the point at which it is opened: the dose necessary for colonization of the intestine.

The question of how many bacteria of a particular strain are necessary for colonization and how many are live in the product will become more important for all probiotic dairy drink makers in the future. Consumers are gaining awareness of these issues – partly as a result of Yakult's marketing – and, if the health benefits are to be credible, one day all manufacturers will have to provide this kind of guarantee.

In 1994, when Yakult opened its European production facility in The Netherlands, the probiotic, daily-dose little bottle was an entirely new concept in Europe. There was no shortage of people willing to express scepticism about Yakult's chances of success – objections included that the concept was too Asian or too medicinal for European consumers; that consumers would not want to be tied to a daily-dose product; and that consumers would not buy a product for digestive health.

Yakult was launched first in The Netherlands and expanded into Belgium and the UK, before being introduced to Germany and France in 1999. In the UK, where some of the greatest scepticism about Yakult was expressed, and where Yakult itself thought it would have difficulty in convincing consumers, the product achieved national distribution in mainstream supermarkets two years ahead of plan and now sells 180,000 bottles a day – making it a US$27 million brand in the UK alone (in comparison to total European sales of US$80m). But the limit to growth in the UK is still a long way from being reached by Yakult. In Australia, where Yakult has been around for five years, two per cent of the population drink Yakult every day – in the UK the proportion is still only 0.25 per cent. The Australian experience indicates that the UK is likely to go the same way – that would be 1.2 million bottles a day, worth around £150 million (US$240 million) in sales.

Most of Europe's largest dairy companies took some time to wake up to the success of Yakult, but in recent years many have been busily introducing similar products throughout Europe. Danone was first on the trail, bringing out its Actimel probiotic in a daily-dose package similar to Yakult's, and is Yakult's nearest competitor. Actimel is a yogurt drink fermented with a strain of *L. casei* called *L. casei Immunitass*, isolated by Danone's own Daniel Carasso Research Centre. First put on the market in Belgium in 1994, Actimel is now available in Spain, France, Germany, the UK and Ireland. The format is a 100 g bottle, although a slightly smaller 62.5 ml bottle can be found in France. Danone claimed to sell around one million bottles a day of Actimel in 1999, making it the continent's largest functional brand. This level of consumption would give Actimel annual sales in the region of US$160 million.

Other lookalikes, spurred by the success of Yakult and Actimel, have followed. Nestlé, for example, although late into the market, took its successful LC1 probiotic eating yogurt and repackaged it into a daily-dose drinking yogurt in an 80 ml bottle, branded LC1 Go!. LC1's active ingredient is *Lactobacillus johnsonii La1*, discovered and patented by Nestlé's Research Centre in Lausanne, Switzerland, with the Pasteur Institute in Paris, after four years of research. LC1 Go, however, has significantly under-performed Actimel and Yakult in the key markets of Germany, France and the UK – although it has done well in Spain and Italy. Even in Switzerland, Nestlé's home market, where it has no competition from Danone or Yakult, LC1 Go! is an under-performer, with a share of the little-bottle market of just ten per cent. One reason put forward is that the Nestlé name is more closely associated in many consumers' minds with desserts than with healthy-eating products. A more likely explanation is that Nestlé is still struggling to find ways of putting healthfulness at the core of its marketing.

The battle of the little bottles, however, has escalated beyond a contest just between Yakult, Danone and Nestlé. In France for example, can be found two lookalikes from the Lactalis dairy group – B'A Force Vitalité and B'A Force Equilibre, both fermented milk drinks in 67.5 ml bottles containing bifidobacteria and soluble fibres – they are in effect synbiotics – one fortified with vitamin C, the other with calcium. Another example is Finland's Valio, the company behind the world's best-researched power probiotic, *Lactobacillus Goldin & Gorbach*, which has also launched a daily-dose product containing LGG in its home market under the line 'bacteria to the rescue'. LGG can also be found in Emmi's Aktifit, a daily-dose product on the Swiss market.

Novartis has become the latest entrant into the daily-dose category in Europe with its Oclea brand, and has created a slightly different proposition – the prebiotic fruit juice as opposed to the probiotic fermented milk – but still building on the daily-dose category. Launched in Switzerland at the end of 1999, the Oclea 'morning drink' capitalizes on the intrinsically healthy image of fruit juice. As an approach it is a far cry from an undifferentiated lookalike, such as Nestlé's LC1 Go!

In summary, a science-based Japanese company, has given birth to an entirely new product category in Europe. As well as creating the European market for little-bottle probiotic fermented milk drinks, Yakult has successfully demonstrated that Europe's consumers are ready to accept the blurring of the boundaries between food and medicine implied in the functional foods revolution. But a key characteristic of Yakult is their commitment to the long term, and we see them setting the marketing and sophisticated communications pace for many years to come.

Case Study 2: Small Is Beautiful for Sweden

Like Yakult's early origins, two small research-based Swedish companies, BioGaia Biologics and Probi AB, are staking their future on developing their own brands of 'power LAB'. The efforts these two companies are making to research and commercialize probiotic lactic acid bacteria illustrate how innovative science-based companies can internationalize a functional ingredient and be at the cutting edge of the functional foods revolution.

Biogaia biologics: using a probiotic to create premium brands

Since its foundation in 1990, BioGaia Biologics has invested over US$20 million in research and development for their probiotic LAB *Lactobacillus reuteri*; secured worldwide patents on its discoveries; and listed its shares on the Swedish stock market. Now the company is capitalizing on its investment with the roll-out of *Lactobacillus reuteri* through a global licensing deal with Chr. Hansen, the Danish-based ingredients group and the world's largest supplier of probiotic and other dairy cultures. Chr. Hansen, which has operations in more than 40 countries will be responsible for the worldwide marketing of *Reuteri*, leaving BioGaia free to focus on research and development.

BioGaia gives its licensees exclusivity for an application – for example, in yogurts – in a region or country. This means that they can focus on building their market position knowing that no one else will be able to offer consumers *L. reuteri* in competing propositions. The licensee can concentrate on communicating the health benefits of *L. reuteri*, building consumer awareness of these and turning its product range into a premium brand.

BioGaia's network of licensees already includes companies such as America's highly innovative Stoneyfield Farms, a company with a strong reputation for its ethical approach to business, which has set about communicating the health benefits of *L. reuteri* to American consumers – who have little knowledge of probiotic bacteria. As an illustration of the global nature of the business, other licensees for the use of *L. reuteri* in dairy products include Switzerland's ToniLait, Finland's Ingman Foods and Japan's Chichiyasu Milk.

Probi – aiming to be the 'first in probiotics'

Probi AB, whose corporate motto is 'first in probiotics', filed the first patent for their 'power LAB' *Lactobacillus plantarum 299v* in 1992. Probi was floated on the Stockholm stock exchange in May 1998 – since when the company's share price has more than doubled.

Probi LAB technology, *L. plantarum 299v*, is fermented with oats to form an oat-based 'soup' from which products can be developed (and is, therefore, not a dairy-based product, but an oat product!).

To commercialize its science Probi has entered into a partnership with Skanemejerier, the largest dairy group in Sweden's Skane region, with manufacturing, marketing, branding and distribution skills. Skanemejerier has committed itself to building up a portfolio of value-added functional food brands and also has a financial stake in Probi.

The first product containing *L. plantarum 299v* was launched in Sweden in 1994. Called ProViva it was a genuine innovation, the first probiotic juice-based drink to appear on the European market, and it rapidly became a success. Total consumption in the whole of Sweden – which has a population of 8.1 million – is around 9 million litres. ProViva is currently on sale in Poland, Germany, Switzerland, Denmark, The Netherlands, Belgium, the UK, Sweden and Finland. It carries an on-pack statement about its benefits – that it is a drink 'with nutritious active bacteria culture to put your stomach in balance' and 'contains oats'.

In 1999, in another innovation, Sia Glass AB, a Swedish ice cream producer, launched what is probably the first-ever commercially-available probiotic ice cream, using *L. plantarum 299v*, outside Japan. Developments such as these demonstrate how probiotics can be incorporated into the diet in a wide variety of regular foods, not just products like yogurts.

Passionate about probiotics!

Probi's chief executive officer, Kaj Vareman, is passionate about the future potential of probiotics and Probi's uncompromising focus on probiotic research. In an interview in 1999, before his retirement, he said:

> *'Bacteria are of vital importance to our well-being. They break down the food we consume and help convert it into energy. They help strengthen our immune system. When we take antibiotics to deal with problems caused by pathogenic bacteria, we also attack the good bacteria, which make up the majority of bacteria in our system. Restoring the balance of good bacteria after taking antibiotics is essential to the maintenance of health – 30 per cent of Swedes have problems with intestinal health, and there are similar proportions of people in other countries such as Germany, Spain and Poland. Probiotics address a real problem for many people.*
>
> *'We are also seeing antibiotics become less effective as the pathogenic bacteria get their own resistance. In the future, I think, probiotics will become increasingly*

important as a preventative measure – by taking them frequently you can reduce your risk of infections, which antibiotics may one day perhaps not be able to cure as they do now.

'Also, around the world healthcare costs are sky-rocketing. Sweden's healthcare bill, for example, has leapt from SEK 3 billion to SEK 28 billion since 1979. Prevention will be of increasing interest in the future to help control the growth in healthcare costs. We stand at the threshold of a new age for probiotics and I think the real explosion of interest has yet to come.'

Case Study 3: European Developments in Yogurts with Health Claims

A large number of probiotic, prebiotic and synbiotic yogurts have been introduced in different European markets (see Table 9.1), with varying degrees of success. In this section we briefly describe a number of products in The Netherlands, France and Denmark, which are typical of many European markets.

The Netherlands

In The Netherlands Campina Melkunie, Europe's fifth largest multinational dairy group dominates the healthy yogurt market with its Vifit, Fysiq and Goede Morgen! brands. The Fysiq brand is a fatless synbiotic yogurt focusing on cholesterol-lowering rather than gut health, and with an on-pack claim which states that it 'contributes to a healthy cholesterol level...the effect is strengthened by the presence of dietary fibre'. It is probably unique in Europe in being the only functional product to have had its health claim legally challenged and tested in court. Because of the weight of scientific evidence presented by Campina about the functional components which contribute to the cholesterol-lowering effect – probiotic *Lactobacillus acidophilus Gilliland* and the prebiotic inulin – the company won and Fysiq, with its cholesterol-lowering claim, continues to develop as a successful brand.

France

The French probiotic yogurt market came into existence thanks to Morinaga Milk, Japan's third largest dairy group and one of the world's top 20 dairy companies. In the early 1980s Morinaga bought St Hubert, a small French dairy producer, and in 1986 St Hubert launched B'A, a yogurt containing *bifidobacterium* and *acidophilus* LAB. It was immediately successful, there being nothing comparable on the market at the time. Danone, who saw their dominant position in the French yogurt market challenged by B'A then introduced a competing probiotic, Danone Bio, which soon came to dominate the probiotic sector with 70 to 80 per cent market share, although it did not make any particular health claims. From sales of just 1500 t at its inception in 1986, the French market grew to about 75,000 t by 1994 – the year it was transformed by the entry of Nestle's LC1 yogurt.

LC1 was first launched onto the French market in 1994, backed by a US$7.95 million promotion, and with the claim that it reinforces the body's natural defence mechanisms. It was an immediate success and had a dramatic impact on the French probiotic market and on Danone Bio, whose market share fell. LC1 succeeded immediately for three reasons:

1 Its packaging was more striking and innovative than anything then on the market and this appealed to the French consumer;
2 The launch was backed by a massive promotional campaign, costing almost US$8 million in the first year;
3 Most importantly, it was the very first time that a yogurt had been marketed in France on the basis of a health claim and it was the first time a company had based advertisements on a scientifically established and documented claim. It was an entirely new and unique point of differentiation from a well-known and trusted company and thus it was successful.

Between 1994 and 1996 the probiotic yogurt segment grew at around 15 per cent per annum, driven by the combined spending on marketing by the three leading brands of FFR 100 million (US$16 million). Nestlé achieved an 11 per cent share of the sector within a year from launch. By the end of 1997 the French probiotic yogurt market stood at 100,000 kg, equivalent to 12 per cent of the total French market by volume, and 18 per cent by value, reflecting the high price premium which probiotics could command at that time.

Nestlé rapidly introduced LC1 to Belgium in 1994, Spain, Portugal, Switzerland, Italy and Germany in 1995 and the UK in 1996. Since then Nestlé has taken the LC1 brand to Australia (where it is one of the two dominant probiotic yogurt brands), Brazil and elsewhere. The most recent move, in early 2000, was to launch LC1 in the US not as a yogurt, but as a probiotic powder packaged and sold as a dietary supplement.

Denmark

In our final marketing example, we consider the fate of Gaio, an early functional food produced by MD Foods. A yogurt, it was sold on a cholesterol-lowering platform, rather than gut health, and, through the use of high-profile advertising and strong health claims, set itself apart from other functional products. Gaio became a big success in Denmark, but in Sweden and the UK it failed.

MD Foods, Denmark's largest dairy company, with annual sales of about US$3.6 billion was among the pioneers in developing functional food. The company's strategy is to expand its business internationally by developing brands and its branded presence at the major retailers by offering foods with positive health benefits. Its merger with Sweden's Arla, announced in 1999, sees it become Europe's largest dairy group.

MD Foods gave birth to the Danish probiotic market in 1984 when it launched Cultura, one of Europe's first *Lactobacillus acidophilus* and *Bifidus* products. However, MD Foods' flagship functional brand is Gaio yogurt, which was launched in Denmark in 1993. Gaio's health benefits – it carries a cholesterol-lowering claim – were attributed to the *Causido* culture it contained. Clinical trials, carried out at the University Hospital in Aarhus, Denmark, had shown that Gaio reduced blood cholesterol levels. Just one year after launch Gaio had captured a 15 per cent share, despite a 70 per cent price premium over regular yogurt and by 1997 it had become the biggest yogurt brand in Denmark. Spurred on by their Danish success, MD Foods decided to launch Gaio in the UK.

Gaio's health claim under scrutiny

The story of how Gaio failed in the UK amid a welter of criticism about its health claim, is a lesson in the importance of ensuring that a claim can be defended by a body of scientific evidence. Launched in the UK in June 1995 and supported by a US$5.7 million promotional budget, Gaio's marketing used the same health claim as in Denmark: 'Gaio with Causido culture actively lowers your cholesterol level when eaten regularly as part of a balanced, low-fat diet.'

Gaio's claim resulted in complaints from the Food Commission to the ASA, MD Foods having failed to obtain prior clearance for its print advertisements from the ASA. The Food Commission said that the advertisements should not be used as the cholesterol-lowering effect of the product was not 'sufficiently proven' adding that a graph on the packaging exaggerated the product's beneficial effect and a comparison with other yogurts was misleading. It was a major setback for Gaio when the ASA ruled that its health claim was not sufficiently founded, fuelling further media and consumer group criticism. MD Foods later issued a statement claiming that the ASA had at best 'misinterpreted or, at worst, ignored' the company's independent research and expert advice.

Before launching Gaio, MD Foods had gone to great lengths to ensure that its marketing messages would meet with the approval of regulators and enforcement authorities. This included putting the evidence before the MAFF Acting Committee on Novel Products (a voluntary process in the UK) which gave the all-clear on grounds of safety. In addition, the clinical research consultants of the British Advertising Clearance Centre (BACC), the advertising industry's regulatory authority on TV advertising, checked the company's research on Gaio and approved it as valid to support TV advertising. Thus while the BACC had cleared Gaio's TV advertisments, the ASA, upheld complaints against Gaio's print advertising. The product was withdrawn from the UK market in January 1997 due to poor sales.

In Denmark, Gaio remains not only the largest probiotic brand but the largest yogurt brand and MD Foods has invested heavily in clinical research to provide further weight to Gaio's claim.

Conclusions: Lessons from the European Experience

In this chapter we have described the creation of a functional food market in Europe based on probiotic lactic acid bacteria. A key feature of this market is the success of true innovation and the emergence of a completely new category – the daily-dose drink concept – based on the proposition of gut health. In 1990 in most European countries such a market was either modest or non-existent. Today it is one of the most dynamic and lucrative functional food markets in all the main European countries, indeed Danone has called it, 'the fastest-growing segment in the fresh dairy industry worldwide' (*New Nutrition Business*, April 2000).

American food companies are missing out on one of the most lucrative areas of functional food developments – namely products for the health of the human gut. Compared to Europe and Japan the US has to be seen as the world's laggards when it comes to this aspect of food and health marketing. What is more, European and Japanese companies are looking to extend their expertise in this area and are eagerly eyeing the North American market. America will not be immune to the probiotic revolution or the battle of the little bottles. Danone clearly has plans to build Actimel into a major international brand to rival Yakult and has begun with the introduction of Actimel into US test markets. Yakult, too, has designs on the US and started test-marketing their product in California in November 1999.

Chapter 10

Functional Food Strategies – Lessons from the Revolution

We have identified seven basic business strategies, which are being adopted with varying degrees of success, in the fight for the success of functional foods. These are:

1. The functional foods make-over;
2. The fortified first-mover;
3. New product substitution;
4. Creating a new category;
5. Incremental business in existing markets;
6. Category substitution;
7. Leveraging hidden nutritional assets.

This chapter provides a short overview of each strategy – and how effective they are. We also look at two of the key issues in determining the success or failure of most functional brands – pricing and marketing communications. But to understand the strategies we must first understand the industry context within which these strategies are evolving.

The Quest for Growth

Finding a strategy which will deliver competitive advantage and growth is one of the greatest challenges faced by managers in the food and drink industry. In this quest a who's who of food and drink companies are turning to functional food in a bid to create truly differentiated, high value-added business.

The reasons for the increasing interest in functional food are not hard to find. The reality of the modern food industry is an unending struggle for profit margins in often unresponsive markets. Competition in products aimed at the lower end of the market is fierce, with profits stripped back to the minimum. Competition in products aimed at higher income consumers is no less severe, though with greater opportunities for brand differentiation and growth.

To the dynamics of competition in fairly mature markets has been added the pressures of ever-growing retailer power, a result of the gathering pace of retail consolidation. Multiple retailers have become dominant in Western food economies, their growth over the last 20 years occurring at the expense of other types of retailer. In the UK 60 per cent of the food retailing market is controlled by four companies; in Germany five supermarkets have control of two-thirds of the market, while in Australia three companies have 80 per cent of national food sales. This dominance has grown out of their ability to establish large outlets and use information technology (IT) and logistics to improve efficiency and lower their costs. However, once dominance has been achieved, continued growth along these lines becomes more difficult, forcing national retailers to become increasingly international and even global in the search for scale economies and enhanced buying power with which to drive down suppliers' prices and, most importantly, to produce continuously increasing shareholder returns.

Meanwhile the shape of food markets is changing, with an increasing trend towards out-of-home consumption – in the US 50 per cent of the 'food dollar' is already spent on food service. In the UK the figure is almost 40 per cent and it is rising across Europe, with The Netherlands, Spain and France at around 30 per cent. This presents a new challenge for food marketers – how to maintain sales of brands in the face of this trend.

Shareholder pressure on food producers is no less severe than for retailers. Stockholders and fund managers demand ever-higher returns and financial markets are quick to mark down the stocks of companies who fail to meet growth expectations.

In response to these pressures, companies are locked into an almost continual process of restructuring – an endless, restless quest to boost productivity, cut costs and add value: factories are closed, production is consolidated in fewer and fewer plants, processes are streamlined and under-performing brands are dropped, and workers are laid off or made redundant. For example, in February 2000 Coca-Cola announced it was cutting 20 per cent of its workforce – 6000 jobs – as part of its global restructuring.

To maintain shareholder interest, companies must constantly reinvent themselves, often in a game of corporate musical chairs, acquiring businesses which reinforce their core competencies and disposing of non-core activities in a ceaseless quest for higher margins, larger scale economies and higher market share.

As a result the global food-processing industry is rapidly consolidating in the hands of fewer companies whose common goal is to be the most successful in their core businesses – everywhere. Danone, for

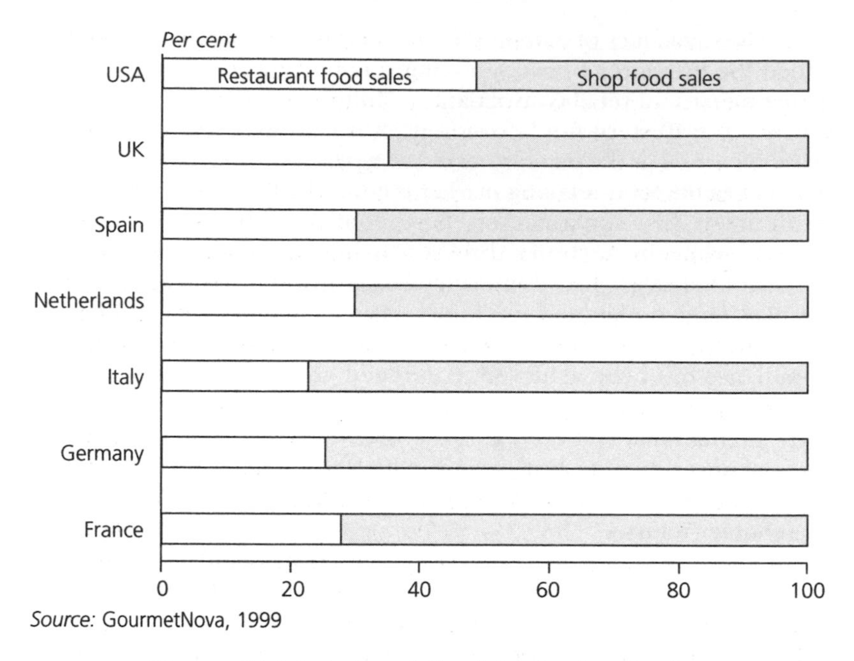

Figure 10.1 *Retail and Food-Service Food Sales*

example, is focused on three core markets – bottled water, biscuits and dairy products – and is pursuing an aggressive acquisition-based strategy in order to achieve leadership in each. To take just one example, in late 1999 Danone became the second largest player in America's $4.3bn bottled water market, the world's largest and fastest-growing, lifting its share from 6.8 to 13.9 per cent with a single acquisition. On a global scale, only Nestlé is now bigger in bottled water.

But once all the avenues to cost reduction and expansion by acquisition have been explored this leaves only new product innovation as a potential engine of growth is left. And this is where the food industry's ambitions hit the buffers. Innovation is not a core competence of the food and drink industries and is only found in a very few companies.

The Challenge of Innovation

Innovation in food and drink lags behind all other sectors, according to research by Price Waterhouse Coopers (reported in *Financial Times Food Business*, October 2000). Price Waterhouse Coopers found that consumer packaged goods (CPG) companies earn an average 34 per cent of their turnover from new products and services. Food and drink companies, by contrast, earn just 21 per cent of their turnover from new products – well below the CPG sector average and the all-sector average of 38 per cent. Top performing innovators, say Price Waterhouse Coopers , are generating 75 per cent of their turnover from products and services introduced in the last five years.

A joint study by Ernst & Young and A C Nielsen, (1999), examined the performance of 24,543 new products in 32 categories – of which half were food and drink – launched in a 13-month period from June 1996 to June 1997 in six European countries. Innovative products, the study found, accounted for a negligible 1.4 per cent of new product activity in the countries studied.

Tellingly, 'lookalike' products accounted for 76.7 per cent of new product launches and these had a massive failure rate. In France, for example, out of 2977 lookalikes launched, just 4.5 per cent succeeded (defined as gaining at least 50 per cent distribution) and almost 75 per cent were dead or almost dead after one year. The implication is that massive amounts of marketing and product development resources are being wasted.

In North America the situation is little better than in Europe. Of 14,519 new food and beverage introductions recorded in the US in 1999 by Marketing Intelligence Service, just 879 (6 per cent) were innovative.

That innovation is lower in food and drink than for other industries is perhaps not surprising, given how little food companies devote to research and development, as Table 10.1 illustrates.

One of the obstacles to innovation in the food industry is that it is historically a low-technology business and true differentiation is very difficult to achieve. Barriers to entry are low, innovations are difficult to patent-protect and product features – tastes, ingredients, packaging designs – are relatively easily and quickly copied by competitors,

Table 10.1 *Comparison of Food, Pharmaceutical and Electronics Industries Research and Development to Value-Added Ratio*

	Electronics	*Pharmaceutical*	*Food*
Denmark	12.7	18.0	1.2
France	n/a	31.7	1.6
Germany	15.6	16.0	1.2
Japan	18.4	13.0	2.0
Sweden	33.2	41.2	1.8
UK	22.9	22.6	1.0

Source: Grunert, Aarhus Business School, Denmark, 1998

hence the proliferation of 'me-too' products identified by the Ernst & Young and A C Nielsen study (1999).

Against this background it is hardly surprising that functional foods have become the focus of excitement for the world's food industry. The functional foods revolution offers the enticing possibility of truly innovative and differentiated products – potentially on a scale never before seen in the food industry. A golden scenario often invoked within the industry is one of products differentiated by a scientifically-proven health claim, perhaps using (as in the case of Benecol) an ingredient or process which is patent-protected; which will generate incremental sales, rather than cannibalizing existing products; and which will command premium prices. But is anyone actually getting to this 'promised land'? What strategies are companies actually following? Which have been successful? And which have failed?

Functional Food Marketing Strategies

In the following sections we identify what we see as seven main marketing strategies being pursued by international food companies in their quest for market success in the functional foods revolution.

Strategy 1: the functional food make-over

Many companies are, in fact, not pursuing a strategy of strong product differentiation. Given the limited experience most food marketers have with innovative products, the low level of corporate spending on research and development and the low-risk approach of a somewhat 'conservative' industry, that perhaps is not surprising. In fact, the single most common strategy among companies trying to boost their business with a health proposition is to turn to what we call the 'functional food make-over'. It is a 'me-too' and 'least-risk' strategy.

The make-over is about taking existing brands and fortifying them with additional vitamins and minerals. There is nothing new about that, as fortification has a long history in the food industry. The term 'fortified foods' meant – as it still means to many nutritionists – those produced with added vitamins or minerals to 'compensate for potential deficiencies in the diet of the entire population' (International Food Information Council, 2000).

One of the most recent and best-known examples of this policy was the US government's decision to impose mandatory fortification of all cereal grain products – such as flour and pasta – with folic acid (a B-group vitamin). The move was based on well-established evidence of the link between folic acid intake and reduced risk of neural tube defects – malformations of the brain or spinal cord in newborn children. It is aimed at benefiting women of child-bearing age who, it is believed, will not otherwise get enough folic acid from their diet. And there are many other examples of mandatory fortification. In the UK, for example, vitamin E is added to margarine and for many years Denmark required the fortification of milk with vitamin D.

The functional food make-over, by contrast, is about food and drink marketers turning to the large-scale fortification of everyday products with vitamins and minerals, not to compensate for widespread dietary deficiencies but as a selling proposition to consumers. It is a product strategy which is sometimes termed 'enrichment' and there are some who argue that a fortified food does not constitute a functional food. But, as the definitions of the term functional food set out in Chapter 1 illustrate, any food which is offering a benefit beyond basic nutrition is functional.

In the eyes of the consumer, a fortified food is just another type of food marketed as having health benefits. And although fortified foods are currently targeted at consumers who want to remain healthy, as the disease-fighting abilities of some vitamins become better understood – for example the growing evidence for a link between folic acid and reduced risk of heart disease – the boundary between health and disease will blur.

Nowhere is the failure of the functional food make-over as a strategy more clearly illustrated than in the breakfast cereal aisles of supermarkets, where vitamin and mineral fortification is now generic to the category and has ceased to be a point of differentiation.

In 1999, in a bid to halt the slide in its US breakfast cereals business, Kellogg turned to a product repositioning which combined two of the hottest topics in the food industry – the 'pester power of kids' and consumers' growing interest in health. The result, launched in the spring of 1999, was K-Sentials, a range of breakfast cereals with an increased vitamin and mineral content, marketed as 'healthy for kids'. Backed by a US$20 million blitz involving TV and print advertising, and health professional communications, K-Sentials was launched under the slogan 'K-Sentials for growth'. Never was a marketing slogan more sincerely meant – K-Sentials, it was hoped, would revive flagging breakfast cereal sales.

Many Wall Street analysts were taken in by Kellogg's description of K-Sentials as an 'innovative initiative', one Deutsche Bank analyst going so far as to describe Kellogg's stock as a 'table-pounding buy' and saying that 'new products' like K-Sentials could return Kellogg's sales to growth (reported in *Financial Times Food Business*, April 1999).

Unfortunately, K-Sentials wasn't new. It was the relaunch of a range of existing brands which between them accounted for about 40 per cent of Kellogg's US breakfast cereals business. Nor had Kellogg created any serious point of differentiation – the increased levels of vitamins and minerals in K-Sentials were only at the same levels already found in some of General Mills' well-established brands, such as Cheerios (reported in *New Nutrition Business*, May 1999).

By the end of 1999 Kellogg had culled its advertising campaign for K-Sentials; it was reported in the December issue of *Advertising Age* that K-Sentials had failed to deliver the expected boost to sales. The make-over had failed to stop the workings of the breakfast cereal market's fundamental dynamics. For years Kellogg had tried to position its cereals as premium products while rival General Mills had relied on the one characteristic of breakfast cereal which has been most responsible for its popularity for almost a century – its relatively low price (Bruce and Crawford, 1995). Kellogg built premium pricing, but failed to create any significant point of differentiation thus allowing cheaper General Mills and supermarket brands to eat up market share.

The result was a slide in Kellogg's value market share in the US from 42 per cent in 1988 to 35 per cent by 1995. The K-Sentials make-over failed to halt this slide and by the end of 1999 Kellogg lost its market leadership for the first time in its 100-year history, its value share slipping to 30.9 per cent, just behind General Mills with 32 per cent (Information Resources Inc, personal communication, 1999).

Kellogg's experience with the inadequacy of the functional food make-over strategy was not confined to the US. In the UK, Europe's largest breakfast cereals market – the British have a higher per capita consumption even than Americans – some brands were 'super-charged' with extra vitamins and minerals just as their prices were cut by 10 per cent.

The problem with fortification is that being a technically easy path to follow, it will always be copied by competitors and is unlikely to be a sustainable source of differentiation. Indeed the food and drink industries' low rate of innovation and high rate of 'me-toos' makes this outcome almost inevitable. A classic example is the German fruit drink market, which had few vitamin A, C, and E fortified products five years ago. The point of difference these drinks had has now been lost and dozens of ACE-vitamin-fortified 'me-too' products are launched every year in Germany.

There is also research which indicates that consumers may not be very motivated by fortified products. A study by the polling organization (BMRB) on behalf of the market research organization Mintel, which surveyed 1052 UK adults, found that a lowly 25 per cent thought that fortified products were beneficial to their health (with no difference between men and women). Only 27 per cent of those surveyed actually read the nutritional information on product packaging and 37 per cent said they did not notice whether a product was fortified. And although Mintel found that consumer perception of fortified foods is not negative – only 11 per cent regarded them as a waste of money – 43 per cent said they thought they got enough vitamins and minerals from a balanced diet, a challenge to many food marketers' preconceptions about fortification.

The functional food make-over delivers no incremental business, no premium pricing, no differentiation, no innovation and is a strategy unlikely to realize the dream of successful functional foods.

Strategy 2: the fortified first mover

Although we have said that vitamin and mineral fortification alone are not enough to build a differentiated brand, there is a caveat. Fortification can be a source of competitive advantage in categories in which there are no major fortified brands already. In such categories the first company to create one can potentially gain significant first-mover advantage. An example is Procter & Gamble's European launch of its Sunny Delight five per cent fruit juice drink. Long established in the US but new to Europe, Sunny Delight was introduced in the UK for the first time in April 1998. Procter and Gamble backed their launch with marketing running into tens of millions of dollars, an investment which dwarfed anything existing category players had previously done, and by December 1998 Sunny Delight had already sold 180 million litres and become one of the UK's top-ten soft drinks brands (Tate & Lyle, 1999).

Procter & Gamble were new entrants in the juice category in the UK; Sunny Delight was a new brand; it was the first significant fortified juice drink brand and it was priced competitively with similar juices and with colas. The fact that it contains fruit juice gives it a healthy image with parents anxious to reduce their children's consumption of colas and its added vitamins appeal to parents conscious of the need to provide a nutritious diet for their children. In fact over 50 per cent of Sunny Delight is consumed by children, according to research company Taylor Nelson Sofres' Family Food Panel (reported in *New Nutrition Business*, March 2000). Described as having 'materially altered' the UK soft drinks landscape, it contributed to a 50 per cent volume growth in the UK fruit drinks category in 1998.

Sunny Delight's competitors in the juice market are defending their positions by turning to the functional food make-over. Drinks group Britvic, for example, fortified their dilutable juices with vitamins at the end of 1999 – at a stroke making 33.9 per cent (Britvic's market share) of the UK's $655 million dilutable juice market 'functional'. Soon, what was a point of difference for Procter & Gamble will become a common feature of the whole juice drinks market.

The essence of the fortified first-mover strategy is the launch of a new product with a proposition – for example, health from vitamins – that is new in the category. A constraining feature is that this strategy will tend to be most effective only for companies with pockets deep enough to commit the marketing needed to make a big impact very quickly and seize a large market share, and it is a strategy open only to one company in a category – the first mover.

Strategy 3: the new product substitution strategy

Substitution – a term made popular among corporate strategists by Harvard Business School guru Michael Porter – refers to the practice of entering completely new markets or categories with new products which consumers, it is hoped, will buy as a substitute for their existing brands. It is the strategy being pursued by Novartis in its bid to become one of the world's leading players in functional food.

Novartis, the Swiss-based multinational consumer healthcare and pharmaceutical group, spent 1999 selling off its 'non-core' food and drink businesses, ranging from Italian sugar-confectionery makers to Sweden's Wasa crispbread brand, retaining well-known European brands such as Wander, Ovaltine and Isostar. At the end of the year Novartis revealed that it was entering totally new categories with the launch of the Aviva range of functional foods, a move described at the company's annual results presentation as 'a significant step toward our goal of defining the new functional food category in the marketplace'.

Positioned as having heart, health and digestive benefits with each product carrying a claim, the Aviva range is a 'complete family' of truly innovative functional foods, competing in the breakfast cereal, cereal bar, biscuit, hot drink and juice drink categories. In only one of these – hot drinks – does Novartis have existing business. Its entry with premium-priced foods with strong health messages leaves it with no risk of cannibalizing existing business. A justifiable fear of brand cannibalization is holding back many companies from plunging into the functional food marketplace – and since consumers will substitute some of their existing purchases for Novartis' brands, the only losers will be the existing category players. As a Novartis spokesman put it: 'Unlike the people in those categories already, we've got nothing to lose.' Novartis' strategy is bold and pioneering: as a new player in its selected categories it is potentially in a winner-takes-all position.

The substitution strategy was also adopted by McNeil when launching Benecol internationally. McNeil had no existing business in margarine, a business which is 'ex-growth', with sales of yellow fats in slow long-term decline around the world as consumers move to healthier fats. In Finland, for example, the market stood at 44,200 tonnes in 1996 and had declined to 43,200 tonnes by 1999. Nor did McNeil have business in the other categories in which Benecol products were launched – salad dressings, yogurt or cheese spreads, for example. Had Benecol been as successful as McNeil planned, it is likely to have been at the expense of brands already competing in those categories.

There is a downside to the substitution strategy: it is high risk. Creating new brands costs millions in promotional spending and can take years – indeed Novartis are talking of a five to seven year time frame for making Aviva successful, as Raisio now is of Benecol. It is a strategy that requires deep pockets, massive resources, long-term commitment and steady nerves. Moreover it is a strategy which, while it may create value for the company following it, may not create any category growth.

In August 2000, Norvartis lost nerve and pulled the plug on the Aviva range in the UK following poor performance. Aviva continued to be sold in both Austria and Switzerland. After just six months in the UK Novartis went back to the drawing board to re-think its strategy. Novartis admitted that they found the UK environment a lot tougher than they expected and said that they did not get their message across quickly and simply enough. Another problem the company had in the UK was in creating a 'nutritional shelf' in supermarkets for the Aviva range; they were unable to find the same location and stick to it in UK supermarkets.

There is also the question of competitive response. In the case of Benecol, Unilever was able to quickly bring out, in the US, a competing product, so undermining the 'uniqueness' of the Benecol proposition. Competitors can bring out 'me-toos' which are lower-priced, so undermining any premium pricing strategy. They can also narrow the gap between the health proposition of new products and their existing brands – as we explained in Chapter 2, the Benecol case study, Unilever's existing Flora brand uses a heart-health position built up by their marketing over 20 years and carries a cholesterol-lowering statement which may be as attractive to consumers as the Benecol proposition – and Flora is at a price point competitive with other spreads and far below that of Benecol.

Strategy 4: creating a new category – and incremental business

To create an entirely new category based around an innovative product is a very rare achievement in the food and drink industries. To successfully develop such a new category into a global business is rarer still. But this is precisely what Yakult has done with their creation of the little bottle daily-dose probiotic drink category, targeted at digestive health (see Chapters 6 and 9).

This strategy, as invented and pioneered by Yakult, is both the simplest to describe and probably the toughest to implement. Its foundation lies on the innovation, in 1955, of a completely new and original product concept. After establishing the concept's popularity in its Japanese home market the company set about rolling its brand out around the world – in one size, one packaging design, with one message, in a uniform style of marketing communications and a refusal to do a private label – in effect, a refusal to embrace the market customization and flexibility in the face of powerful retailers which is the gospel of much of today's food marketing thinking.

In most countries which Yakult has entered it has been first-to-market and in each of them Yakult has come as an innovation every time. Its European launch in 1995, was a classic example, with many European dairy companies dismissing the newcomer's prospects and many retailers uncertain as to whether they should stock this unusual drink.

Yakult's success is the reward for a coherent and consistent business model which offers:

- An innovative – even revolutionary – product at its core;
- A model of global market-building proven successful over 30 years and followed in every new country in which Yakult is launched;
- A long-term vision underpinned by a coherent philosophy and belief that the company's goal is to enhance human health (which we examine more fully in Chapters 7, 9 and 11);
- A simple proposition which relates to real consumer need – around 25 per cent of people have some kind of digestive problem in any three-month period, according to Yakult – and which has also been identified by Novartis as one of the three top health concerns;
- A long-term vision which recognizes that building a functional food brand with a loyal consumer base is a five to ten year business, with profitability arriving only over a similar time horizon; and
- Linked to this long-term view is a long-term commitment to invest heavily in consumer education as the strongest way of creating brand awareness and loyalty.

And this model is applied – with little variation – around the world.

Some of Yakult's success is almost certainly because their concept is – even after 45 years – still highly innovative in almost every new market which the company enters. When Yakult entered Europe it faced competitors with limited experience of innovation or of the nutrition marketing approach which characterizes its own communications. Despite years of growth around the world, its entry and success in Europe still came as a surprise.

The innovation-based new category strategy also holds appeal for retailers looking for sales growth. Substitution is a very attractive route for many companies aiming to enter the functional food 'fray', but it is about replacing existing brands, in effect cannibalizing part of the category. And while it may increase a category's value by persuading consumers to trade up to higher-value products, it offers only limited possibilities of real incremental business from drawing in new consumers. By contrast, creating a new category – although a high-risk strategy – means real incremental business, which is exactly the experience of European retailers who stock Yakult.

Competitors looking to follow with 'me-toos' will have to decide whether they want to play and how – they will have to learn a new set of market dynamics. Their lack of experience – compared with Yakult's 45 years with essentially the same business model – will tell in the marketplace.

Another innovator and new category creator is New Zealand Dairy Foods, New Zealand's largest dairy company, which gave birth to the daily-dose category there when it launched its Metchnikoff Symbio drink – a synbiotic, using both probiotic bacteria and prebiotic fibres. It is the first and so far only such daily-dose product on the New Zealand market. The brand is backed by a strong nutrition marketing and education approach, modelled on Yakult, and is developing into a profitable niche business.

Yakult is probably one of the very few companies to have fulfilled the potential of the functional food dream – a dream of truly innovative, differentiated, premium-priced products leading to the creation of new markets. This strategy is one of the most exciting of the functional foods revolution and it is one which can be adopted successfully by other companies in other categories.

Another good example of new category creation is the US energy-bar market. It barely existed in 1995, when the Balance Bar Company had sales of just US$1 million. But by 1999 the market was valued at an astonishing US$500 million, and Balance Bar, the second biggest player in the business, had sales of US$100 million. The creators of this market were not global corporations but small, privately held businesses who embraced innovation and risk but, importantly, also had a coherent philosophy: energy bars were designed by people interested in outdoor pursuits to be eaten as a portable energy source and a healthy alternative to chocolate bars, by people taking part in outdoor pursuits The rewards for having the courage to create a new category were massive – in early 2000 Balance Bar with its 90 employees sold out to Kraft for US$268 million and Power Bar, number one in the market, sold out to Nestlé for an estimated US$375 million (reported in *New Nutrition Business*, June 1999).

As Yakult and the US energy bar companies have shown, new category creation is a high-risk, but high-reward opportunity. Our belief is that it can be repeated, if the desire to innovate and take a long-term view is followed and if a company is willing to be bold and challenge established ideas. How many corporate managements will risk such boldness?

Strategy 5: incremental sales in existing markets

In practice very few companies will have the resources at their disposal to enter completely new categories – like Novartis – or to create new categories – like Yakult.

Compared to operating within the known parameters of existing markets, either approach is high risk and most companies will be reluctant to move into areas in which they have little or no experience. Many companies will be constrained by their limited research and development resources. They may have the capacity to innovate within known markets, but not to make the huge increase in investment necessary to innovate in new categories. The obvious way forward for many is to find ways of creating new brands to compete within existing categories with the aims listed below, so resulting in an increase in the value of the category:

- Getting existing consumers to trade up to higher-value brands;
- Drawing more consumers to the category;
- Getting existing consumers to consume more.

The least desirable situation is to introduce new propositions which simply cannibalize existing business, producing little or no net gain. Europe's probiotic-eating yogurt market, as we explained in Chapter 9, is a good example of how some companies have gone about following this strategy. In France the launch of Nestlé's LC1 yogurt in 1994 expanded the country's small probiotic category and contributed to overall yogurt market growth. There was a similar effect in Germany where the volume of the German probiotic yogurt market increased from just under two million kilograms in 1995 to almost six million kilograms in 1997, with a corresponding increase in value from about DM 10,000 to DM 32,000 (quoted in *Milch Marketing*, April 1997)

The development of probiotics as an engine of growth for the overall yogurt market can also be seen at work in Australia, where the probiotic yogurt market has shown annual growth rates of around 20 per cent in recent years, led by Nestlé's LC1 and Paul's Dairy (part of Parmalat) with Vaalia, a yogurt using the LGG bacteria (see Chapter 8).

In Australia, these 'new brands' offer consumers a new proposition with simple-to-understand health benefits. In the early days of the market's development they can grow alongside existing brands and contribute to category growth. There will come a point of maturity, however, when the brands such as these could begin to eat into sales of existing business. This is the nub of question – can functional brands become mass market brands without cannibalizing existing business, or should they remain as niche propositions? If companies want to maintain premium pricing and incremental business then they must pursue a niche strategy – to go to mass-market will inevitably result in margins being eroded and the point of difference of functional products being lost.

Strategy 6: category substitution

The idea behind this strategy is very simple – it is about taking the 'health' selling proposition of a competing category and applying it to yours, using health to attract more users to your category at the expense of another category. One of the best examples is the positioning of calcium-fortified juice as an alternative to milk. For example, juice manufacturer Tropicana, a division of PepsiCo, has taken the unique proposition of milk – that it is a good source of calcium – and applied it to orange juice.

Fruit juice has an intrinsically healthy image with consumers. Milk, by contrast, has an image problem – consumers associate it with fat and cholesterol. The New Zealand Dairy Advisory Board, for example, has said that in surveys consumers consistently think that liquid milk's fat content is around 20 per cent when in fact full-fat milk has just a 4 per cent fat content. Research from New Zealand indicates that teenage girls in particular are reluctant to drink milk because of its 'fat' image (personal communication to authors).

Tropicana Pure Premium Calcium is a 100 per cent orange juice fortified to provide as much calcium as the same amount of semi-skimmed milk in a 250 ml glass. It is clearly targeted as a substitute for milk – 'As much calcium as milk!' reads the UK packaging copy –

and to drive the point home also depicts side-by-side an image of a glass of orange juice and one of milk and the words '38 per cent RDA'. Tropicana also emphasizes that it is 'an excellent source of calcium for people who dislike milk or are lactose intolerant' and 'contains a more absorbable form of calcium than other sources'. Tropicana Calcium effectively becomes a better source of calcium than milk.

Tropicana Calcium is likely to develop the fruit juice category at the expense of milk. It is priced competitively with Tropicana's other 100 per cent orange juice products – in Europe at a five to ten per cent premium (on what is already a premium-priced category), and in the US at parity. The calcium-fortified proposition is not unique to Tropicana – its competitor Minute Maid (owned by Coca-Cola) also has a calcium-fortified 100 per cent juice, but the competitive advantage of these calcium-fortified brands may be sustained for far longer than products which undergo a functional food make-over. Consumers are already switching from milk – sales are in long-term decline – to fruit juice because of its intrinsically healthier image. Tropicana's calcium message reinforces their motivation to do so.

Calcium-fortified orange juice has become a highly successful strategy capturing 20 per cent of the chilled orange juice market in the US. In 1999 the total value of sales of chilled juices and drinks in the US was US$4278 million of which calcium-fortified juice totalled US$583 million and Tropicana US$253 million (according to Chicago-based market research company Information Resources Inc, 2000). It is worth noting that the success of this strategy is that it is based on the selection of the ingredient. A particular ingredient's health benefits allow a particular marketing position to be adopted. By choosing calcium, juice makers are able to position themselves against a competing drink category and build on the high consumer awareness of the the benefits of calcium in relation to bone health.

Strategy 7: leveraging hidden nutritional assets

Generic claims are not a popular idea within some quarters of the food industry; they sit at the opposite end of the spectrum from product specific claims. When many products can carry the same claim the opportunities for differentiation, growth and premium pricing disappear. But the developing success of soy – particularly when compared to the troubles of Benecol – suggests that in fact the route of generic claims may be a valid alternative strategy for achieving sales growth – though not premium pricing.

Sales of soy products enjoyed a boom in the US in 1999 as the message about its health benefits reached a wider consumer audience

– total sales reached US$418.7 million. Of these the sales of tofu reached a total of US$46 million (a 19 per cent increase when compared with 1998); soy milk sales totalled US$77.4 million (an increase of 85 per cent); meatless hotdogs, burgers and deli meats totalled US$176.2 million (a growth of 26 per cent) and soy-based energy bars totalled US$82 million with a remarkable growth of 93 per cent (data from market research company SPINS/A C Nielsen, 1999)

The health position of soy was confirmed in October 1999 when the FDA ruled – as had been expected – that foods containing a minimum of 6.25g of soy protein per serving could carry a health claim. The claim has two similar approved forms, one of which reads:

> *'25 grams of soy protein a day, as part of a diet low in saturated fat and cholesterol, may reduce the risk of heart disease. A serving of (name of food) supplies X grams of soy protein.'*

Any company whose product meets the FDA's criteria can put this generic claim on their product.

Growing consumer awareness of the health benefits of soy is the fruit of a massive communications campaign by the major producers, ADM and PTI (part of DuPont). A survey by the United Soybean Board in 1999 found that 71 per cent of consumers recognized it as 'very healthy', compared to only 59 per cent in 1997. More than 40 per cent of those surveyed were aware of the various health benefits claimed for soy which, as well as heart health, include reduced risk of osteoporosis and alleviated symptoms of the menopause.

Against this background soy products have moved from the whole-food store, where they were traditionally sold, into mainstream supermarkets and are now being adopted into the diet by an increasing proportion of Americans. The scope for growth in soy consumption is massive – already 24 per cent of Americans say they use soy products once a week or more, compared to only 15 per cent in 1998.

The major soy suppliers have also invested heavily in helping food manufacturers find ways to use soy protein as an ingredient in a wide range of foods – including breakfast cereals, vegeburgers, pasta, muffins and bread – at levels which qualify for the FDA-approved claim, so allowing consumers to easily include soy in their diet. It is a strategy which could result in the heart-health claim appearing on a very wide array of products.

These products, however, will not be highly differentiated from one another and they will not be able to earn high price premiums, if they earn any premium at all. The evidence, in fact, is that no-premium pricing is already becoming established. One store check carried out for this book (*New Nutrition Business*, February 2000) found a leading brand of soy milk carrying the FDA's heart-health claim priced below that of another brand without the heart health claim!

Overall sales of all soy products have grown and are likely to go on growing as a result of the whole category benefiting by association with the generic claim. Soy products are also likely to be net gainers at the expense of other categories – with some consumers substituting soy for dairy milk, for example, or soyaburgers for beefburgers, an outcome with something in common with the strategy of whole category substitution.

The soy case study illustrates a direction in which the whole functional foods revolution could evolve as nutrition science reveals more and more about the intrinsic health-giving properties of food and food components, enabling an ever-widening range of foods to make health claims.

General Mills' whole grain strategy also fits under the generic claim heading. General Mills in the US (and in Europe in partnership with Nestlé) has begun to make extensive use of the FDA-approved generic heart health claim for whole grain on long-established brands such as Cheerios. There has been no change or addition to Cheerios in order to benefit from the claim – it is rather a case of General Mills leveraging off the brand's intrinsic and (until now) hidden nutritional assets. Significantly, General Mills' brands with the claim are priced comparably with regular breakfast cereals (see Figures 10.2 and 10.3).

Developments like these challenge the idea of functional foods as foods priced at high premiums with product-specific health claims. Indeed it would be difficult to see how this strategy can remain viable if the same benefit is available generically from a wide range of foods which are priced to similar levels as regular products. Indeed there is a certain irony that heart health, the message behind Benecol's cholesterol-lowering claim and the message which it was hoped would enable it to earn five times the price of regular products, is one of the first claims to have become generic and can be found on an increasing number of whole wheatgerm, oat and soy-based products – ranging from soy milk to burgers to breakfast cereals.

Strategies Summary

For convenience, Table 10.2 summarizes the seven marketing strategies currently driving functional food markets:

Table 10.2 *Summary of strategies*

	Examples of companies using it	*Examples of categories*
Strategy 1: Functional foods make-over	Kellogg	Breakfast cereals
Strategy 2: Fortified first mover	P&G	Fruit juice drinks
Strategy 3: New product substitution	Novartis	Biscuits, cereal bars, hot drinks, breakfast cereals, juice drinks
Strategy 4: Creating a new category	Yakult	Probiotic fermented milk drinks
Strategy 5: Incremental business in existing markets	Nestlé, MD Foods and many other European dairy groups	Probiotic yoghurts
Strategy 6: Category substitution	Tropicana	Fruit Juice
Strategy 7: Leveraging hidden nutritional assests	Heinz Soy industry Cranberry juice producers	Lycopene and tomatoes Soy Cranberry juice

Strategic Partnerships

The search for 'hit' ingredients, which can be used as the basis for successful functional products has resulted in a rash of strategic partnerships between global food and drink giants and technology-focused ingredients companies. The former bring global distribution, marketing and branding skills 'to the party', offering the innovative partner the opportunity to access markets and achieve income streams which would be out of the reach of such small companies by themselves. The latter, often very small, with sales of just a few million dollars, offer the ability to innovate, a skill which as we have already seen is in short supply in the research and development departments of many large corporations. The following list is a 'snapshot' of strategic alliances and licensing deals in functional foods completed between January 1999 and March 2000:

- Swedish research and development company BioGaia Biologics forms global alliance with Chr. Hansen, world's largest supplier of probiotic cultures to market its *L. reuteri* probiotic;
- GalaGen and American Institutional Products Science Company licenses rights for a functional dairy beverage for healthcare;
- GalaGen and Novartis – gives Novartis global rights for its colostrum-based Proventra natural immune components in certain nutrition product applications;
- Forbes Medi-Tech and Novartis – Swiss pharmaceutical giant Novartis becomes global licensee for Canadian research and development company's cholesterol-lowering plant sterols;
- NutriPharma and Del Monte – Norwegian research and development company NutriPharma licenses its cholesterol-lowering soya-based ingredient to juice drink maker Del Monte for Europe and Middle East;
- NutriPharma and Sinnove Finden – NutriPharma's soy ingredient licensed to Norwegian dairy group Sinnove Finden for use in yogurts;
- Galaxy and Tropicana – small US soy ingredient company forms alliance with Tropicana to develop cholesterol-lowering soy-based 'smoothies';
- Clover Healthcare and Merck – German pharmaceutical company Merck becomes global marketing partner for innovative Australian company's omega–3 fatty acids;
- Science Foundation and Merck becomes global marketing partner for patent-protected naturally-occurring folates;
- Scotia and General Mills – Scottish pharmaceutical group Scotia gives distribution licence for its satiety technology to General Mills in the US.

The typically worldwide character of these alliances underlines the truly global nature of the functional foods revolution. The worldwide alliance between BioGaia Biologics, the science-based Swedish research company, which has worldwide patents on *Lactobacillus reuteri*, and ingredients giant Chr. Hansen, discussed in Chapter 9, is just one of many examples of this trend.

Partnerships are also becoming a common theme in finished foods. In early 2000 Novartis and Quaker, for example, announced the creation of a new joint venture company, Altus, equally owned and managed by the two companies, to develop and market functional food in North America. McNeil was chosen as the worldwide marketing partner for Benecol because, it was believed, it brought skills and resources in global branding and marketing which Benecol, operating primarily in Scandinavia, did not possess. For food giants like Danone, partnerships offer the ability to gain access to categories and markets which would otherwise require millions of dollars in product development and marketing costs. To take one example, Danone has taken a 20 per cent stake in Lifeway, a US$10 million Chicago-based maker of branded kefir (fermented dairy) products, an alliance which sees Lifeway gaining from access to Danone's distribution to supermarkets across the US.

Pricing Strategies

To achieve premium pricing for functional brands is, for many companies, the holy grail of the functional foods revolution, and one of the key objectives which justifies the many millions of dollars invested in the product development process. For any marketer in any industry, setting the price of any new product can be one of the most challenging professional tasks they face. Correctly pricing functional foods is proving to be particularly difficult.

Put simplistically, in most markets – although there are a few exceptions – a highly priced product will tend to perform as a niche brand, while a product priced closer to regular products will yield higher volumes. One of the key issues in pricing is whether companies are trying to build a niche or a mainstream brand. Interestingly, some of the highest-profile functional food activity to date has seen companies aiming for mainstream-type volume, but at the same time pitching products at niche-type market prices, the rationale being that consumers are prepared to pay substantial premiums for health. The example of Benecol throws doubt on the validity of this idea. McNeil set out to market Benecol as a mass-market brand but positioned it at five times the price of regular products. It was not surprising that sales of Benecol were what might be expected from a niche product.

Finding the maximum premium which can be charged before a product becomes niche rather than mainstream is the key issue. The difficulties were set out by Al Piergallini, president and chief executive officer of Novartis Consumer Health Worldwide, in an interview with *New Nutrition Business* (February 2000), when he explained that he had already cut the prices of Aviva functional products by 15 to 20 per cent in just three months after their launch, adding:

> *'and if it proves necessary we'll reduce prices again until we find the level which gives a premium but also draws in sales. We've got to get the balance right between premium pricing and building volume.'*

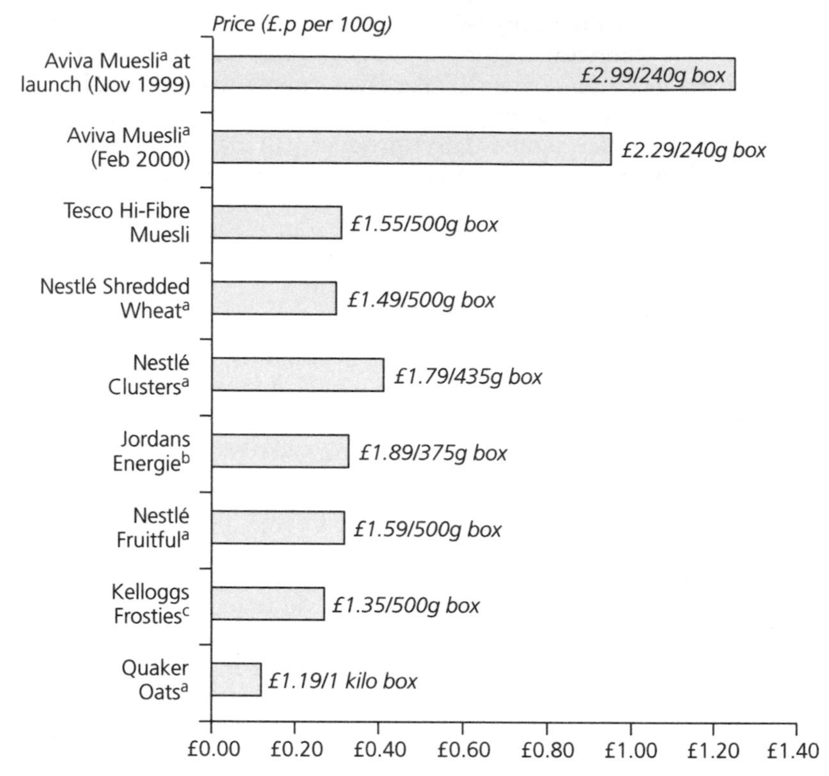

Notes: (a) carries 'Heart Healthy' claim; (b) claim 'A natural source of energy'; (c) 'Now with added calcium'
Source: Store checks carried out at west London branches of Tesco, April 2000

Figure 10.2 *Pricing comparisons, UK breakfast cereals, April 2000*

Part of the answer to the pricing issue can be found by looking at how differentiated a product is and at how easily consumers can compare one product with another. A heart-health claim, for example, is still perceived by some marketers as providing strong differentiation and therefore a justification for charging a high price premium. But in fact this difference is being undermined all the time, as we explained earlier in this chapter, by the growing proliferation of generic heart-health claims on regular-priced products. This competitive issue confronts Novartis, for example, whose UK Aviva breakfast cereal carried a heart-health claim and a significant price premium (200 per cent over comparable cereals), putting it under pressure from the proliferation of breakfast cereals carrying similar 'scientifically proven' claims, but priced just like regular products (see Figures 10.2 and

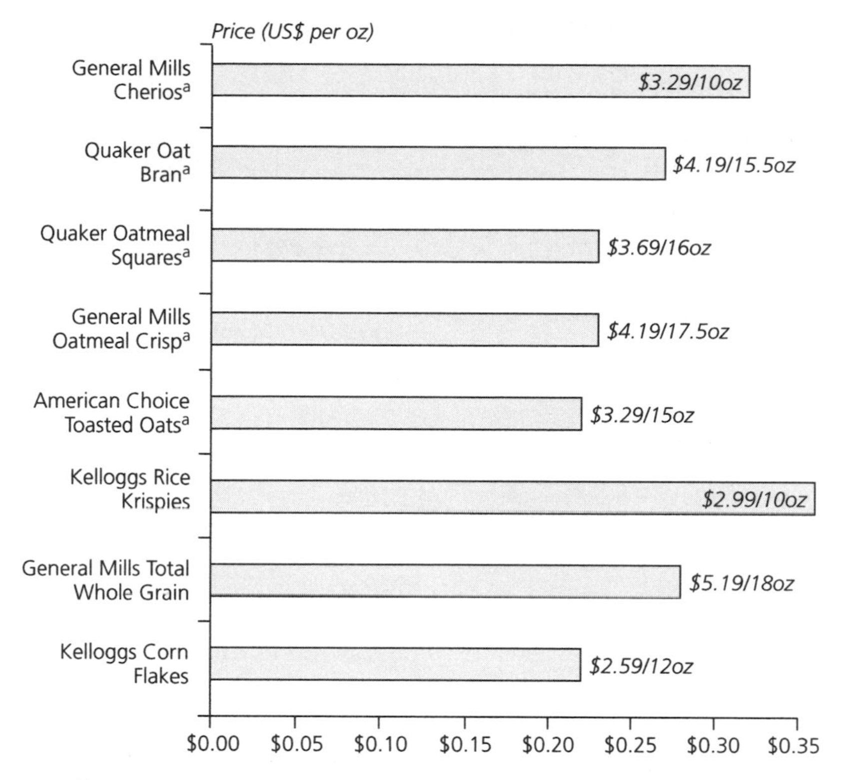

Note: (a) carries 'Heart Healthy' campaign
Source: Store check carried out at Farmer Jacks Supermarket, Detroit, Michigan, April 2000

Figure 10.3 ***Pricing comparisons, US breakfast cereals, April 2000***

10.3). Faced with the rapid loss of its point of difference it may develop as a niche rather than as a high-volume brand.

In some new categories high prices are being successfully maintained – such as the daily-dose little bottle market, where the price of the finished product represents a huge level of value-added over the cost of raw materials. An admittedly crude, but nonetheless illuminating, example which illustrates the point is to compare the retail price of a litre of semi-skimmed liquid milk, which retails in the UK for 30 pence per litre , with the price of Yakult – milk which has been fermented with probiotic bacteria and water added – which retails at an impressive £5.47 per litre equivalent.

One of the advantages of creating a new category like this is that consumers cannot easily make direct price comparisons with similar

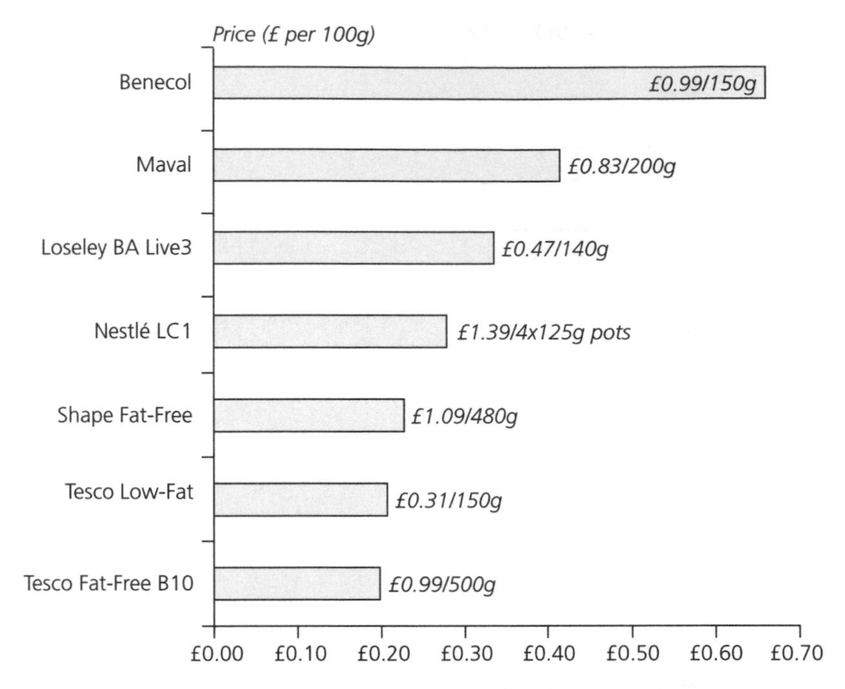

Source: Store checks carried out at west London branches of Tesco, April 2000

Figure 10.4 *Pricing comparisons, UK yoghurt market, April 2000*

products. The first mover can set a price point for the whole category and, provided that 'me-too' entrants use this as a benchmark for their own pricing, then consumers can only compare with similar – and similarly priced products. The effects of this can clearly be seen at work in the daily-dose category, where Nestlé's LC1 Go!, for example, a Yakult 'me-too', is priced only 10 per cent below Yakult.

Europe's probiotic yogurt market is notable for some successful products achieving good price premiums. As Figure 10.4 shows, Nestlé LC1 probiotic yogurt, carrying a claim, is priced at around a 40 per cent premium to regular yogurts in the UK, although its market share is fairly modest. Figure 10.5 shows a similar situation in Denmark, where Gaio, a probiotic yogurt with a cholesterol-lowering claim, and Cultura, an *acidophilus* and *bifidus* yogurt with the claim 'the way to a healthy stomach' are both sold at a 40 per cent premium to regular yogurts – a similar premium to that charged for organic products.

However, these price premiums are a long way from the 100 per cent plus aimed for by Aviva and Benecol. Indeed it is questionable

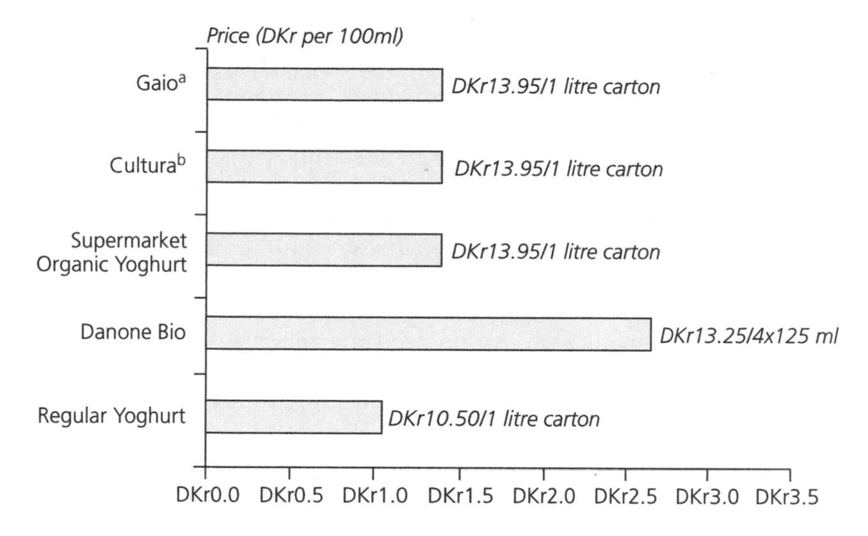

Source: Store checks carried out at west London branches of Tesco, April 2000

Figure 10.5 *Pricing comparisons, UK yoghurt market, April 2000*

how sustainable 100 per cent premiums will be in many markets. Maval, for example, a probiotic yogurt produced by Sweden's Skane Dairy and sold in the UK and Sweden, had a unique selling proposition. Made with an ingredient called Olibra, licensed from pharmaceutical group Scotia, it carried the claim that it had the 'ability to generate a longer feeling of fullness', a claim supported by clinical trials which demonstrated that Maval produced a feeling of satiety, causing people to consume less fat at their next meal. Olibra has also been licensed in the US by General Mills, which has yet to launch an Olibra-based productat the time of writing. A skilful PR campaign backing Maval's launch attracted a huge amount of media attention to a unique proposition which seemed set to be a certain winner with weight-conscious consumers. Yet Maval's price – a 100 per cent premium to regular yogurts – deterred many consumers and in early 2000 it was withdrawn from the market.

One Day All Food Will Be Functional

The growth in generic claims, often appearing on products with pricing comparable with 'regular' products, threatens to undermine many premium-price positions. It is a process that will accelerate as the increasing pace of nutrition research identifies health benefits from a wide range of everyday foods. One day, perhaps, 'all foods will be functional', whether through an added ingredient to provide the health benefit or through 'discovering' and leveraging the intrinsic benefits of the food, as in the case of soy – many may carry a health claim. How then will high price premiums for specific brands be maintained? How will such products compete at premium prices when the supermarket is filled with soy, whole grain or oat beta-glucan-based foods, all priced like regular foods and carrying a scientifically proven, FDA-approved heart-healthy claim? One answer could be for companies to develop products aimed at specific medical conditions. Heart disease might seem to be one attractive example, but prevention by dietary means is already rapidly becoming an everyday proposition. Rather, product developers could focus on products for niche conditions or for use in customized diets.

One development that is already under way – and which will become more important in the future – is that consumers will be willing to 'trade up' from categories with a 'less healthy' image to categories with higher prices but a strong health proposition, for example, substituting soy milk or calcium-fortified orange juice for cow's milk. In the US this means moving up from, for example, dairy milk priced at US$1.09 a quart to soy milk at US$1.89 a quart.

New Foods, New Communications

The issue of health claims is at the centre of the debate around functional food. The regulatory complexities of this issue are covered in Chapter 5; this section focuses on the role of claims in marketing communications.

A health claim – and ideally a product or ingredient-specific claim – is viewed by many in the food industry as a powerful, even an ideal, way of differentiating a functional food and a strong way of communicating a product's benefits. It is, however, far from certain that a claim is as important as widely believed – the evidence on the usefulness of health claims *per se* is, at best, mixed.

Product-specific claims may not hold the strong appeal for consumers that many marketers imagine. For example, the NCC (1997) found that health claims confused consumers more than they helped them, made little sense to shoppers, and many were suspicious of longer and more complex claims (see Chapter 5).

More recently, in a Mori National Opinion Poll (NOP) in 1999, on behalf of the Mintel market research organization, Mori interviewed 922 adults and produced what Mintel called 'a damning assessment' with 77 per cent of those surveyed believing that 'a lot of health claims made by food manufacturers about their products are misleading' (Mintel, April 1999).

In fact, far from claims being essential for functional food success, some companies have, through skilful marketing and communications, proved that it is possible to build a highly successful health positioning, widely recognized by consumers, without using any claim. It is worth noting, for example, that the boom in US sales of soya products in 1999 occurred mostly prior to the approval of a heart-health claim by the FDA in October of that year. This was driven by a skilful communications campaign masterminded by the soy producers which successfully raised consumer awareness of soy's benefits. The success of General Mills with their 'whole grains' strategy is in their marketing and advertising skills, not the FDA approved health claim itself.

The Promotion of Lycopene: the Power of Effective Communications Strategies

One of the best examples is the promotion of lycopene, firstly by the tomato industry and more recently just by Heinz. Lycopene is an antioxidant which gives tomatoes their colour. Studies have identified a link between consumption of tomato products and a reduced risk of prostate cancer. There is no health claim for lycopene approved by the FDA (or any other regulator) and Heinz makes no claims for the benefits of lycopene on its tomato products. In many countries the neck labels of Heinz tomato ketchup bottles simply bear the statement 'a source of lycopene'. The lycopene message itself gets to consumers through a sophisticated communications strategy.

In 1996 the Tomato Research Council (TRC) was formed, under the auspices of the National Food Processors Association, to educate industry and consumers on the importance of the health benefits of processed tomatoes. The Council was funded by Hunt-Wesson Company, Campbell Soup Company and HJ Heinz, all three of which are major processors of tomato products such as tomato paste, tomato sauce and tomato ketchup. Each company reportedly contributed in excess of a million dollars towards this effort. The TRC worked closely with the public relations agency Aronow & Pollock Communications Inc to coordinate publicity on tomato products. The TRC does not sponsor lycopene research, although some of it is sponsored by its member companies. In 1996 the TRC appointed a Scientific Advisory Panel which contributes to the credibility of this programme.

In March 1997 the TRC and the American Health Foundation together sponsored the international scientific symposium on the role of lycopene and tomato products in disease prevention in New York City. It included leading scientific researchers investigating lycopene. A press conference at the symposium resulted in an article on lycopene in the *New York Times* (Brody, 1997).

The TRC newsletter *The Tomato Research Digest* highlights research into processed tomato products and lycopene. It provides answers to common questions, and 'Tomato Tidbits' which gives interesting statistics on tomato products. It is a well-designed public relations product to provide scientific information in an easy format that the media can readily use.

The TRC actively publicized the latest scientific results on lycopene, by issuing press releases highlighting specific research and successfully elevated lycopene as a 'hot topic' in functional foods. Through its scientific symposia, the TRC stimulated communication among scientific experts conducting the research, possibly resulting in greater collaboration and/or effectiveness of research efforts.

At the end of 1998 the partners behind the TRC decided to go their separate ways, but Heinz, which processes two million tons of tomatoes annually, has picked up the lycopene health promotion mantel, setting up the Lycopene Education Project with the goal of continuing to promote the health benefits of tomatoes and lycopene:

> *'We want to share the information with consumers, who are less concerned with the FDA health claim than they are with what they see on TV, in magazines or on the Internet,'*

said Kevin Krail, general manager of functional foods business development for Heinz in an interview in the February 2000 issue of *Advertising Age*.

Heinz has set out to associate the lycopene claim with its flagship brand, first with a controversial US$400,000 print campaign which depicted a bottle of ketchup with the headline, 'Lycopene may help reduce the risk of prostate and cervical cancer' and the Cancer Research Foundation logo. The advertisement ran just once, in January 1999 – it reached eight million consumers and sparked an investigation by the Federal Trade Commission, which looks into advertising claims, but the investigation was closed in part because of the advertising's short run.

The advertisement achieved the aim of stirring interest in lycopene and the Lycopene Education Project, which provides information through a web-site; booklets for dietitians and teaching guides for middle schools. The lycopene campaign extended beyond the US to Canada, Europe, Australia and New Zealand gaining 352 million media 'impressions' in 1999. These efforts, Heinz has said, have helped the company increase its share of the US ketchup category by four percentage points.

Good Communications as the Key to Success

Good communications are increasingly proving to be the key to the success of functional products. The NCC report, referred to earlier, concluded that consumers need information about products to be reinforced by other trusted sources and that claims on labels will not change consumers' behaviour unless the message has been reinforced elsewhere. The challenge for food marketers is to find these convincing sources.

Japan's Yakult also uses a highly sophisticated communications strategy that includes advertising, but rests primarily on an integrated approach to consumer education. The company is emphatic that this has been the prime generator of business for their brand since 1955 – indeed Yakult believes that less than 50 per cent of its sales arise from advertising. In every country in which Yakult is launched its uses approximately the same model of communication – often, and literally, face-to-face with consumers. In Australia, for example, a nutritionist leading a team of five dieticians visits dietetics departments in hospitals, doctors and gastroenterologists, their brief being not to sell Yakult but to raise the awareness of the importance of probiotic bacteria in intestinal health. The company encourages people to visit the Yakult plant – 8500 did in 1998, and these included school parties and women's associations. A similar approach is used in every European country in which Yakult can be found and each year thousands of consumers visit the Yakult factory in The Netherlands to learn about probiotics and health.

Yakult has demonstrated that a far-reaching education campaign about intestinal health, targeted at consumers and health professionals alike, is fundamental to building up enduring credibility and market success for a functional product. One element of this is the frank discussion of the mechanics of intestinal health in its communications materials. The company produces educational literature, such as its consumer-friendly *Guide to the Gut*, a useful guide to stools and the workings of the intestines. And any consumer who walks up to a supermarket shelf carrying Yakult will probably find a shelf-edge dispenser, full of leaflets – 'shelf-talkers' – reiterating the intestinal health message.

A massive and continuous sampling effort also underpins the Yakult communications strategy – over one million Yakult bottles have been sampled in the UK alone since 1996 – and each sampling campaign is treated as an opportunity for Yakult educators to talk to consumers face-to-face about their product.

Heinz and Yakult prove, if proof were needed, that there is a lot more to making a success of a functional product than just a claim.

The Market for the Health Proposition

Functional foods do not exist in isolation; in fact they are just one component of a broad spectrum of food and drink products marketed on a health proposition. This spectrum includes organic foods, natural products, vegetarian foods, 'traditional' healthy-eating-type reduced-fat foods and some everyday foods with intrinsic nutritional advantages, such as the well-publicized benefits of the Mediterranean diet. Functional foods are in competition with all these health propositions for the consumer's attention.

It may seem strange to refer to organic and natural foods, for example, as foods competing on the health proposition, but there is a growing body of evidence that, from the perspective of consumers who are interested in the links between their diet and their health – the target group for functional foods – that this is precisely what is happening.

For example, Dragon, the respected European brand development consultancy with a track record in functional food and drinks, ran a series of focus groups in the UK in 1999 with women aged 20 to 55. Their findings present some clear challenges to marketers. The most startling was that functional foods, defined by Dragon as 'processed foods made with added nutrient or therapeutic ingredients like vitamins or calcium', were viewed by consumers as less healthy than organic or conventional fresh foods, or organic processed foods. Dragon asked consumers 'What type of food would you say is the healthiest, in terms of doing you the most good or least harm?' The response was revealing with the results, ranked in order of importance, being as follows:

1 Processed foods made from organic ingredients;
2 Fresh produce produced using conventional methods;
3 Processed foods made with added nutrients or therapeutic ingredients (like vitamins or calcium);

4 Processed foods made with conventionally produced ingredients;
5 Fresh produce which has been genetically modified to require less pesticide in production;
6 Fresh produce which has been genetically modified to have high levels of nutrients or health-giving ingredients (Dragon, 1999).

Consumers also agreed that organic products had changed their image in recent years from unattractive, worthy products for the 'sandals brigade' to highly aspirational products preferred on the basis of taste – and health! Dragon went on to ask consumers what they thought food retailers' and manufacturers' priorities should be. The answers they gave, ranked in priority, are as follows:

1 Making organic foods cheaper and more available;
2 Reducing the use of pesticides and herbicides in agriculture;
3 Improving animal welfare in farming;
4 Ensuring that the producers of crops in developing countries have good wages and conditions;
5 Introducing higher standards of hygiene to reduce the risk of food poisoning;
6 Keeping the cost of food as low as possible;
7 Using naturally occurring therapeutic materials to provide foods which can reduce the risk of certain diseases;
8 Developing ways of keeping fresh food longer;
9 Developing food which tastes better;
10 Using techniques of genetic modification to develop crops which can reduce the risk of certain diseases (Dragon, 1999).

The results of this survey show that functional foods came in at a lowly seventh place. A separate survey of 2000 shoppers by the Consumers' Association got similar answers, with 80 per cent of those surveyed (only 2 per cent of whom actually bought organic foods regularly) giving health as their primary reason for buying organic and no less than 46 per cent of the sample believing that organic food contains more vitamins and minerals than conventionally produced foods! Results such as these, if consistent and wide-ranging, suggest some fundamental issues relating to food and health (and food production) needed radically reassessing!

Natural and organic products are relevant because of their rapidly accelerating impact on consumer spending. The US in particular is seeing a boom in sales of natural products. It is hard to come by a single definition of natural products but, according to Tom Aarts, vice president of Health Business Partners, a US venture capital company which specializes in backing natural product companies: 'The definition comes from the marketplace – what products the whole food stores will stock. Their definition has become institutionalized.'

The retailer Wild Oats Markets is one such example. It sets out its widely used definition as (*Financial Times Food Business*, December 1999):

- Free of artificial preservatives;
- Free of artificial colours;
- Free of chemical additives;
- Organically grown, wherever possible;
- The least-processed or unadulterated version available;
- Non-irradiated;
- Cruelty-free.

According to ADM, one of the world's largest food groups, sales of such foods in the 'Health and Natural Foods Stores' channel doubled from 1992 to 1997 to US$11.2billion. US industry newsletter *Nutrition Business Journal* says that in all channels sales topped US$19.4 billion in 1998, an 11 per cent growth over 1997. Significantly, mainstream grocery stores accounted for 36 per cent of this total – a sign of the rapid movement of a food category which was once regarded as the preserve of 'health freaks' into the mainstream. In Europe, too, organic foods have moved out of the healthfood store and can now be purchased in most of the largest supermarket chains.

While natural foods are still a niche, accounting for barely 5 per cent of the US food supply, so too are functional foods. According to one estimate, US sales of functional foods were around US$14.7 billion in 1997 – smaller than the natural products market – and in 1999 reached US$17 billion. Another estimate puts European functional food sales in 1999 at about US$2.5 billion – less than Europe's organic foods market, estimated by the USDA at US$4.9 billion in 1998 and forecast to be 40 per cent higher in 1999 (see Table 10.3).

In some countries in Europe the organic concept is having a major impact on the food supply. In Denmark, for example, over 20 per cent of liquid milk consumption is already organic and it is forecast that within a few years around 25 per cent of the country's food supply will be organic! These figures should give the food industry pause for thought. Many of the shoppers for natural and organic products are higher income, often older, health-conscious individuals – the characteristics often said to define functional consumers. There need not, however, be any incompatibility between the functional and organic propositions. Some products are coming on to the market that combine both. To take just one example, the US-based Horizon Organic Dairy has introduced a probiotic, vitamin-fortified organic yogurt.

Table 10.3 ***The European Organic Market***

Country	*Organic retail sales (US$ m)*	*Population (m)*	*Per capita*	*Percent of total sales*	*Estimated growth rate (%)*	*Organic sales imported (%)*
Germany	1.8 bn	81.5	22.0		5–10	50
France	675–725	55.6	12.6	0.5	20	NA
UK	450	58.5	7.7	<1	30–40	70
Switzerland	400	7.26	55.1	2.5	30–40	NA
Denmark	375	5.2	72.1	<3.0	30–40	25
Netherlands	350	15.4	22.7	1	10–15	60
Austria	270	8.0	33.8	2.5	10–15	30
Sweden	125	8.8	14.2	1	30–40	30
US	4.5 bn	270.0	16.7	1	20–25	NA

Source: USDA, *Financial Times Food Business*, November 1999

Of course, natural and organic foods are just one small part of the ever-widening health platform of everyday foods. A surprising amount of nutrition marketing has been going on in connection with long-established brands over many decades. An example from Europe is Flora margarine, a Unilever brand. As we explained in the Benecol case study in Chapter 2, Flora has been marketed on a heart-health proposition for over 20 years. Under the 'Flora Project' Unilever supplies consumers and health professionals with educational materials about heart disease risks (including smoking as well as dietary risk), funds heart research, makes donations to heart-health charities and sponsors conferences and high profile 'health' events, such as the London Marathon. Flora carries a scientifically proven cholesterol-lowering claim on its packaging – and retails for the price of regular margarine!

To this activity is added the effects of the strategy, explained earlier, of leveraging nutritional assets – for example, as used by Heinz with its lycopene promotion and General Mills with its 'whole grain' concept – and emphasizing the intrinsic health benefits of food components, yet pricing and selling them comparably with regular foods and so competing with the idea of premium-priced functional foods.

In summary, consumers are being bombarded with health messages. Moreover, these different propositions are often targeted at the same group of consumers, interested in how their diet can maintain or improve their health.

How to stand out is the challenge facing any new functional food. In fact amid all the noise about health, functional foods are competitively disadvantaged, compared to most rival health propositions, which have their own coherent overarching philosophy.

A Coherent and Clear Philosophy Behind Products with a Health Image

To put it very simply, olive oil, tomatoes and other characteristic elements of the Mediterranean diet benefit from a coherent and very widely accepted healthy image. Vegetarian products are also backed by a coherent philosophy and clear definition with broad appeal – a case could be made that more vegetarian products are consumed by mainstream, meat-eating consumers trying to eat healthily than by committed vegetarians! We have already seen how natural products have benefited from a clear image; image is also proving effective in the organics niche.

The consumer research in the UK by Dragon, quoted earlier, found that organic products had a coherent philosophy and set of values in consumers' minds, which adds emotional appeal and ensures that the term 'organic' communicates immediately and effectively. The integrity of products was seen as guaranteed by the on-pack symbol from the organic certifying body – usually the Soil Association in the UK – with organic packaged goods representing the 'gold standard' in some market sectors.

Coherent philosophies and universally applied standards serve to reassure consumers and can be a powerful way to underpin business. Regulatory authorities can play a key role in this, benefiting both industry and consumers. Sweden, for example, has a well-developed national healthy-eating product symbol, called the 'keyhole symbol'. Products bearing the keyhole symbol are very common and consumer surveys show that the symbol is recognized by a large percentage of consumers. It is used on a wide variety of foods, including meat, spreads, dairy products, breakfast cereals, ready meals, bread and biscuits, provided that they meet precise nutritional criteria. Australia and New Zealand's heart associations, working together with industry, have developed strict criteria – almost a global 'gold standard' – for products that have heart-health benefits. Products that meet these criteria (now numbering 450) can carry the associations' 'tick-mark', which is recognized by 60 per cent of consumers.

There is even one example of coherence in the world of functional food. In Japan the FOSHU symbol – although getting off to a slow start and still scoring low levels of consumer recognition – provides reassurance about the efficacy of products, protecting both consumers and manufacturers who have invested in creating products of proven effectiveness.

Conclusion

In this chapter we have identified and detailed seven distinct marketing approaches using the functional foods revolution as their basis. Some are proving very successful, while others are not at all convincing. In all cases companies are facing a number of major challenges and tough choices if they decide to participate in the functional foods revolution, particularly for long-term and sustained marketing success. We summarize these strategic choices as:

- Functional food make-over versus innovation;
- Cannibalization versus incremental business;
- Regular pricing versus premium pricing;
- Mass market versus niche;
- Targeted at disease versus product for health;
- Claim versus no claim;
- Scientific versus 'tell a great story';
- Advertising versus consumer education;
- Quick hit versus long-term brand building.

There can be little doubt that functional foods – seen from both the manufacturer and consumer perspective – would benefit from the development of a coherent framework. But it would be a mistake to think that a regulatory framework or a symbol for health would alone be sufficient. More important still, for companies and consumers, is that participants in the functional foods revolution develop a coherent philosophy. 'Telling a great story' about functional food is crucially important. The power of creating an image of functional foods and the companies who make them, and communicating it well, is being seized by very few companies. It is a huge opportunity waiting to be tapped and in Chapter 11 we will explain and expand on this point.

Part 5

The Healthful Company™: a Leap Forward

Chapter 11

Becoming a Healthful Company™

As we have shown, at the heart of the functional foods revolution is a simple and completely obvious idea – that food has health-promoting effects in humans.

John Milner, at Penn State University in the US, reminds us that one of the most compelling reasons for the widespread interest in functional food comes from the consistent findings that increasing fruit and vegetable consumption is accompanied by a reduction in the risk of heart disease and cancer (Milner, 1999). Although much remains to be learned, according to Walter Willett based at the Harvard School of Public Health, optimal health can be achieved from a diet that emphasizes a generous intake of vegetables and fruits. He says: 'such plant-enriched diets...can be not only healthy, but interesting and enjoyable as well' (Willett, 1994).

The dynamics of the functional foods revolution, however, show that the business of food and health is far from simple. It is creating unprecedented marketing and product challenges, putting food and nutrition policy into a spin, and rewriting the food legislative rulebook. Yet it is based on two still-to-be-proven assumptions: that the functional foods revolution will produce both healthy people and healthy profits.

In the 1970s nutrition and health appear to have been a limited part of food advertising, but by the late 1980s and into the 1990s, it had become a significant theme (Ippolito, 1999). The functional foods revolution is taking the promotion of nutrition and health to new heights. The food industry's marketing of food, nutrition and health is now taking place globally on an unprecedented scale. Narrow definitions of functional foods – such as those which try to pin the concept only to products making scientifically validated health claims – are wholly inadequate to describe the breadth and scale of this revolution in nutrition science and nutrition marketing.

Although the functional foods revolution is poised to sweep across the modern food economy, it presents difficult and complex issues for the food industry which we have outlined throughout this book. In particular, we see functional food marketing clashing with the culture of the healthy-eating revolution, and of promoting scientific studies assessing the risk of disease reduction in ways that could be construed as giving a false sense of security to the general public. What is more, from our global analysis of market activity it is our assessment that, with a few notable exceptions, functional food business strategies are in danger of falling far short of expectations.

Putting these issues together, and taking into account the new business and consumer environments that are emerging (see below), we offer a radical solution: it is nothing less than a new business model for future foods, a concept we call the 'Healthful Company™'.

The six key characteristics that distinguish The Healthful Company™ are:

1 The health of the consumer is at the core of all activities. Health is absolutely central (a Healthful Company™ adopts a 'holistic' concept of health);
2 A Healthful Company™ adopts a strong 'decency positioning' (what could be a more ethical or socially responsible business than food and health?);
3 A Healthful Company™ builds relationships with consumers and becomes 'emotionally' involved (concepts such as 'respect' and 'humility' are watchwords);
4 A Healthful Company™ moves away from thinking about products to making an 'offer' that is both product and service simultaneously;
5 A Healthful Company™ lives and tells a 'great story';
6 A Healthful Company™ is knowledge-rich (it is here that science, nutrition and technology play their role).

We explain these attributes in more detail below. The Healthful Company™ is not just about individual products or ingredients, it is about a way of being, and of going about business. It moves beyond a fixation on health benefits to the lifestyles and well-being of the people it wishes to serve. It is not about product differentiation, but whole company differentiation. Put simply, where is the Body Shop of the functional food world? Before we discuss in more detail the characteristics of the model Healthful Company™, we first need to outline some of the assumptions that are important to, and inform the concept of, The Healthful Company™ (but are tangential to its day-to-day working).

Can the Functional Foods Revolution Make a Major Contribution to Public Health?

The central premise of the functional foods revolution – that is the production and promotion of foods and beverages with health benefits – once again focuses the spotlight on the role of food and diet in public health policy. What is remarkable, from a food policy perspective, is the way that food companies are adopting the language of public policy in the promotion and marketing of functional food, for example by arguing that cholesterol-lowering products can shift the risk curve for coronary heart disease, thus reducing the incidence of the disease; seriously proposing that functional food will help to reduce the staggering and growing costs of healthcare in the developed world; or suggesting that functional food will enhance and improve the quality of life of a whole generation of 'baby boomers' as they move into old age. None of these assumptions are being evaluated or empirically tested in relation to the marketing of functional food.

The next issue – who will really benefit from functional food – faces the whole food economy, but it would be a glaring omission if we did not draw attention to it in the context of the functional foods revolution. While functional food science is motivated by its potential for public health benefits, the bulk of diet-related disease and illness, together with other health-related problems, is skewed towards people caught in lower socio-economic groups. Yet the benefits of functional food science are, in general, currently geared to those willing and able to pay for premium-priced products. How will functional food translate into health benefits for those most at need?

A concern for some nutritionists is that functional food represents a technical fix that will detract from dietary advice that emphasizes changes to the total diet over time (Lawrence and Raynor, 1988). Another issue is that functional food does not address the basic problems in nutrition, in particular the fact that diet-related disease affects the most disadvantaged sections of society or does not even address the most pressing nutritional needs.

For example, one of the most disturbing health trends in countries such as the US is the prevalence of overweight, affecting more than one-third of the population. Overweight is correlated with, or even a cause of, diet-related illnesses such as heart disease, some cancers, cerebrovascular disease and diabetes mellitus – four of the ten leading causes of death in the US (Bush and Williams, 1999). In particular, diabetes mellitus is increasing at alarming rates. Overweight is also associated with gall bladder disease, osteoarthritis, sleep disorders and psychological stress. The distribution of overweight in the US is more of a problem for women than men, but both income, race and ethnicity are significant predictors of overweight. As Bush and Williams explain, the problem of overweight is not limited to adults only. What they describe as especially alarming is the prevalence of overweight among children and adolescents.

Although age-adjusted death rates from heart disease in the US declined by almost 26 per cent from 1985 to 1996, it is still the leading cause of death for all adult racial and ethnic groups. But again, socio-economic status, racial and ethnic disparities are significant. Death rates are higher with lower incomes, in every category of, sex, race and ethnic groups, and higher among African Americans than among whites.

It is a similar story in the UK with diet-related illness. Professor Phillip James and his colleagues say 'diet affects the health of socially disadvantaged people from the cradle to the grave' (James et al, 1997). In the UK, social class differences in health are seen at all ages, with lower socio-economic groups having greater incidence of premature and low birth-weight babies, heart disease, stroke, and some cancers in adults. Risk factors, including lack of breast feeding, smoking, physical inactivity, obesity, hypertension and poor diet, are clustered in the lower socio-economic groups (James et al, 1997a). Suzi Leather, now deputy chair of the newly formed Food Standards Agency, uses the graphic phrase 'the making of modern malnutrition' in detailing an overview of food poverty in the UK (Leather, 1996).

Finally, while throughout this book we have limited our discussion to the developed world, the problem of diet-related illness and disease is most pressing in the developing world. For example, the United Nations Development Programme (UNDP) calculate that worldwide there are around 800 million undernourished children, with two billion people exhibiting effects of poor diet (UNICEF, 1998). In absolute terms, more people live in poverty today than 20 years ago. About a fifth of the world's population – 1.3 billion people – live on a income of less than US$1 a day (Lang and Heasman, 2001). Jeremy Rifkin (2000) sums up the situation thus:

> *'The reality is that Americans spend more on cosmetics – $8 billion annually – and Europeans on ice cream – $11 billion (in US dollars) – than it would cost to provide basic education, clean water, and sanitation for the 2 billion people in the world who currently go without schooling or even toilets.'*

Michael Gibney and Sean Strain (2000) describe it as deeply shameful that as we enter the new millenium with so many new and exciting concepts of nutrition and health, there is still a legacy of nutritional deficiency in many developing countries. They write:

> *'As the developed world chooses to redefine optimum nutrition, it leaves behind a staggering debt of pathologies of nutritional deficiency [in the developing world] ...now that we are redefining optimal nutrition, we are also redefining the scale of suboptimal nutrition...we will face a new problem of ensuring our new concept of optimal nutrition applies equally well to all people.'*

Challenges such as these can only be addressed in the policy arena. Such 'political nutrition' sits uncomfortably with many nutrition scientists, not least because of its confrontational politics. For example, the history of the healthy-eating revolution is, in part, nothing less than a momentous and bitter battle between public health policy, food industry interests, and state policymakers and regulators. Policy expert Laura Sims, in her book *The Politics of Fat*, describes in considerable detail how both the food industry and federal policy makers in the US have opposed food and nutrition policy changes, despite a strong public health mandate for change (Sims, 1998). She suggests merging agricultural and food production policy concerns with health policy concerns as a possible approach to nutrition and health policy. But, she warns, 'Applying new technological fixes in the form of novel foods is not a strategy that will improve the American diet.'

Functional food science, however, is opening up new policy opportunities. Today a whole range of foodstuffs are being actively, and for the first time, promoted for their health benefits. This includes soy, tomatoes, juice drinks, oats, whole grains, red wine, peanuts and other nuts, a host of dairy products, essential fatty acids, a range of vitamins and minerals, and fruit and vegetables. The scale of activity opens up new public health possibilities for the promotion of good health through diet which have yet to be seized.

There are already, however, notable public policy and food industry collaborations on nutrition promotion which offer lessons for the future. These include a number of 'tick' and 'symbol' schemes between food industry and various heart-health associations, such as the American Heart Association or the work of New Zealand and Australia heart associations that set criteria for products which are recognized as contributing to heart health, but which also meet the guidelines of healthy eating. A further notable example is the '5-A-

Day' programme promoting consumption of fruit and vegetables in the US. These are all cooperative solutions which meet needs of industry, consumers and health and nutrition professionals. The impact of functional food in providing new healthy choices can be no better illustrated than by studying the dairy shelves of supermarkets in countries like The Netherlands or Finland, that have been transformed by functional food (and increasingly organic) options.

Implications for the Functional Foods Revolution

Such developments are giving force to the idea that one day all foods will be functional. The rapid advance of nutrition science has revealed that many foods have health-promoting properties, so progressively undermining the potential of premium-priced products carrying product-specific health claims. These trends are likely to result in supermarket shelves being increasingly filled with products bearing scientifically proven cholesterol-lowering or heart-health claims (which is already happening), but all priced just like any other everyday food in their category. How then is the consumer expected to react to products such as Benecol or Novartis' Aviva which also bear such claims, but are priced at significant premiums?

The essence of these developments is that one of the key strategies of many companies – to achieve strong differentiation of their products by using disease-related health claims – is being rapidly eroded in value. The widely held belief that technology – in the sense of advances in nutrition science – is going to provide differentiation is coming apart under the twin pressures of generic health claims and the rapid diffusion of the technology (such as cholesterol-lowering 'solutions') to a large number of companies. Product differentiation on the basis of technical fixes of single risk factors, such as by how many percentage points a product may or may not lower cholesterol, is looking like a strategy doomed to failure in the long term. It is our view that in an environment of reducing product differentiation, only a Healthful Company™ will be truly differentiated (see below).

Are Health Claims 'Vital'?

The short answer to whether health claims are vital to functional food product success is no. We believe that as the years go by, health claims will be seen as less and less relevant. But by this we are not implying that science is unimportant – on the contrary, we believe the scientific base for food and health can only become more important. But we believe the reliance on the belief that product health claims will transform the market for functional food products may be misplaced for four reasons. First, the level and burden of proof, together with the time needed, will make it unlikely that single companies can achieve product-specific health claims, so other means of marketing and promotion will have to be developed. Second, unless there is new thinking among regulatory authorities along the lines of the Japanese FOSHU system, health claims will be limited and restricted to generic claims, which are open to anyone meeting the relevant criteria. For example, Ovesen (1999) lists some of the drawbacks policy makers see in the development of health claims:

- Fear of disease will be the motivating factor for the purchase of food;
- Dietary information will be entrusted to food producers;
- Foods of high nutritional quality may be substituted by less nutritious foods with health claims;
- May highlight the favourable aspects at the expense of adverse properties;
- Counteract the fact that it is the total diet over an extended period that is essential to health.

Third, generic claims open up the market for all foods to become functional. Companies should not bank on health-claim regulation alone for product success. But there are some exceptions in the short term to this. As General Mills has shown, through brilliant marketing including capitalizing on a generic health claim, much can be achieved. In effect General Mills has used the generic health claim on whole grains to not only leverage its hidden nutritional assets but it has achieved first-mover advantage (see Chapter 7). As the health-claim regimes evolve in Europe and other countries there will be similar one-off marketing opportunities that offer unique chances for product differentiation for those both bold and skilled enough to take advantage of them.

Finally, a raft of health claims on product packaging may become confusing and misleading for consumers. There are mixed research findings on the impact of labelling claims in changing eating behaviour. For example, despite nutrition content claims on some foods in the US, little is known about whether these foods making the healthy-eating claims are replacing similar foods that are not making such claims. Nor is it known who is using claims and to what extent claims are leading to an overall healthier diet (Bush and Williams, 1999).

There is a real dilemma here. How are consumers to know if a product is genuine without some form of regulatory approval? There is still scope for some creative solutions, based on ensuring consumer protection, for developing a system for health messages on food. A good example would be in the area of probiotic products. As we detail in Chapter 9, for probiotics to deliver any sort of health benefit, a certain quantity of live probiotic lactic acid bacteria needs to be present in the product as consumed and the probiotic should be scientifically shown to colonize the lower gut. At face value there would appear to be a strong case, based on consumer protection, to enable those manufacturers who can demonstrate this health effect to be able to say so on packaging to differentiate these products from other 'probiotics' that do not meet the live lactic acid bacteria standards. As keen consumers ourselves of probiotic dairy products we would like to know!

For The Healthful Company™, being able to use regulated health claims is incidental, for every day it will be making its claims to health.

The Six Characteristics of The Healthful Company™

We now turn to look in more detail at the six characteristics of The Healthful Company™. As will be apparent, every food company could become a Healthful Company™, but we focus on the concept as it applies to food, health and the functional foods revolution. The Healthful Company™ is a simple idea and if we are honest, not startlingly original. In fact we use widely available and popular sources to illustrate the thinking behind its characteristics. But what is innovative about the concept of The Healthful Company™ is that it takes the best from futurist thinking about business, companies and markets and applies it uniquely to the business of food and health. And for a concept so simple and obvious as The Healthful Company™ it is surprising so few companies appear to see its potential.

More than anything, The Healthful Company™ means translating ideas about food and health to consumers in ways that engage them emotionally in their own health and well-being through food products and services. It means The Healthful Company™ will have to live and tell its own 'great story' (Jensen, 1999). Such new business thinking is becoming widely disseminated, and is not exclusive to food, but it will play itself out in unique ways as part of the functional foods revolution. Moreover, as the opportunities for single product differentiation are reduced as the functional food revolution unfolds, The Healthful Company™ model will offer, for the few bold enough to take it up, a unique and lasting source of differentiation which will go beyond technology, beyond products and stem from the heart and soul of the company.

The model of The Healthful Company™ is critically important to achieve sustainable, long term business development; capture the trust and commitment of consumers and be credible in the eyes of the world.

It is also about getting in tune with the long term trends shaping a fast- changing business environment; the societies we live in and how we as people will come to view the world. It also recognizes that dietary change and intervention, as we described with the implementation of nutrition policy in Finland, is a long-term process.

We arrived at the radical concept of The Healthful Company™ after analysis of functional food developments throughout the world, asking why, with so much promise in so many cases, a large number of them underperform.

The functional foods revolution aims to change the way most of us in the developed world think about and consume food and drink. Our analysis of current trends, from nutrition policy to regulation and marketing activity, has led us to believe that it will take a new type of company to succeed in functional food. This new company will demonstrate a new approach to the whole issue of food and health and will be a company committed not to getting a hit product within a year or two but to being successful in functional food in the long-run – over 10 or 20 years. It will be necessary for many companies to become Healthful Companies not only for business success, but in order to ensure that consumers, regulators, policy makers and the media take functional foods seriously. Against this background we now look at the detail of The Healthful Company™.

The health of the consumer is at the core of all Healthful Company™ activities

'You're going to get sick and you're going to die.' It's not nice to be reminded of these facts of life all the time. But this is precisely what is happening every day along the supermarket aisles. In much of the functional foods revolution and especially in the US and the UK, functional foods are sold on the back of the fear of disease (wrapped up in statements that refer to risk reduction), and by implication are constant reminders of our mortality. On the other hand, for most people food is, or should be, a source of enjoyment, pleasure and sociability, and it follows naturally that good food and good diet implies good health.

Food, including functional food, should be marketed as something enjoyable that contributes to health and life – not marketed on fear of illness, such as heart disease. This is a core concept of The Healthful Company™ – food is for life, not the prevention of the risk of disease and illness. Consumers today do not merely expect that their food or their environment should not damage their health, but increasingly expect that it should promote their health and well-being. It is a subtle change in emphasis that we see as critical to becoming a Healthful Company™.

At a more theoretical level, this idea captures what is described as the new public health movement or holistic concept of health which is undergoing a renaissance in public health strategy in many countries (McKinlay and Marceau, 2000). This approach focuses on lifestyles and living conditions that determine health status and the factors that promote good health rather than exclusively focusing on the manifestations of ill-health (Lang and Heasman, 2001).

It is a philosophical change in thinking which, put simply, assumes two very different conceptions of health. McKinlay and Marceau (2000) describe these as the 'medical science' (mechanistic) conception of health versus the 'holistic' view of health (also referred to as an 'ecological' approach). The former focuses on the actions and motives of distinct individuals, the latter takes a collectivist approach, with the emphasis on categories (for example, social class, race and ethnicity) or places and social positions in society. The medical science view of health, most prominent in the US healthcare system, focuses on disease and on factors that predispose people to, are associated with, or increase the chances of catching a disease. Thus health becomes seen as a state of non-disease, and disease of people being in a condition of disequilibrium. In contrast a holistic or more collectivist approach, sees the individual as in a condition of 'existential equilibrium' and manifests itself through the lifestyle policies of the new public health, which are of particular interest in Europe.

We are assuming that the new public health or holistic approach to health is the way for the future and that consumers are increasingly viewing their own life situation in the context of a holistic approach to health and well-being. A possibly unique practical example of this type of thinking is New Zealand's Green Prescription scheme. Although developed to help contain spiralling drug costs, physicians are encouraging patients who visit their surgeries to review their lifestyles, diets and exercise regimes, whenever this would be more effective than using drugs. In other words, it is about encouraging people to take a holistic approach to maintaining their health. The boom in alternative medicine and health care is another manifestation of consumers' holistic thinking about health (to such an extent that alternative medicine will cease to be alternative in the near future).

A more holistic approach to heath has the potential to transform the world of drugs and medicine, again in part driven by the needs to reduce costs. For example, several drug companies in the US are already changing their business models from selling drug products alone, to introducing the idea of disease management. Under this model, the drug company agrees to take responsibility for the total treatment of a patient, including disease prevention, patient care and the administration of drugs. The pharmaceutical company Eli Lilly, for example, has singled out five major diseases – diabetes, heart disease, central nervous system disorders, cancer and infectious diseases – for disease management. By shifting focus from selling drugs to servicing patients, companies like Lilly hope to move up the value chain (Rifkin, 2000). The Healthful Company™ will likewise have to shift its thinking from product to service as we discuss further below.

It will be an insuperable challenge for many food companies to put health and consumers first because very often marketing is driven by the physical assets companies own, which in turn dictate what they are willing to sell. A large number of giant food companies are 'dinosaurs' from the age of bulk production, when quantity and the accompanying intensification of production was the context of food supply. It is an irony of the functional foods revolution, which focuses so much on the health benefits of micronutrients in food and their role in chronic disease, that much food processing results in the mass loss of these health-promoting components. For example, in the standard milling of white flour, as much as 60 to 90 per cent of vitamins B–6 and E, folate, and other nutrients are lost (Willett, 1994).

Although we do not address this issue here, this model for food production is now being seriously questioned and its very sustainability becoming more problematic (see for example, Hawken et al, 1999; Lang and Heasman 2001). Like dinosaurs, many of these food companies are powerful and appear very frightening, but there is often not much intelligence driving so much bulk. In an era of rapid change and transition there is much scope for a new life force in the world of food.

A Healthful Company™ adopts decency positioning

The ethical approach of The Healthful Company™ can most easily be explained by an idea set out in 1991 by popular futurist Faith Popcorn, head of the US-based marketing think tank BrainReserve. Her best-selling *Popcorn Report* (1991) identified some key consumer trends shaping society and markets. Many of these, ten years on, have proven to be prescient. One of these trends was the development of what

Popcorn termed 'the Vigilante Consumer'. Popcorn pointed out that the baby-boomers – the oft-cited target market for functional food – are also the protest generation, and that as this generation aged it would flex its muscle in ways which reflect its youth. She cites as an example the ability of the US vigilante consumer to protest and secure the removal of tropical oils (thought at the time to contribute to elevated cholesterol) from cookies in the early 1990s. 'Deep at the heart of the Vigilante Consumer trend is a wish that companies could somehow be more human,' says Popcorn (1991, p73) and argues that increasingly companies will have to act 'to set standards for themselves that meet the standards of the consumer'.

She calls this corporate stance 'decency positioning' arguing that: 'The Decency Positioning is still up for grabs in almost every category.' Importantly, she adds that it is also one of the hardest positions to convince corporations to adopt. Ten years on and the decency positioning is still 'up for grabs' in the functional foods revolution. Few companies have put a truly ethical philosophy of health at the centre of their business practices in the same way that the personal care group The Body Shop, for example, has put the cruelty-free philosophy at the centre of its business practices.

Perhaps through instinct rather than intentional strategic planning, within the food industry the decency positioning has been appropriated by organic and natural foods companies. Their coherent decency philosophies (outlined in Chapter 10) are making a major contribution to help them win the loyalty of a growing number of consumers – and the reward for adopting this positioning has been increasing sales and profits. It is interesting to note that while mainstream food companies have been focused on the disease-fixing properties of functional food and nutraceuticals, the organic and natural foods industry has 'stolen' the food and health mantle. Organic produce is more likely to be identified by consumers as health promoting than most mass-marketed packaged functional food products.

Organic food is an example of consumers taking up a holistic view of health through purchasing behaviour. Although organic production has its roots in ecological concerns it is also about consumers choosing foods which they believe will contribute to their health – as well as the health of the planet (see Chapter 10). It is important to note that environmental health and human health are becoming viewed more and more as indivisible.

The socially responsible consumer is today exercising enormous and growing market power, not just in food. For example, the US-based consultancy group Natural Business, which specializes in the natural products industry, has described this consumer trend as the 'emerging LOHAS' market, LOHAS meaning 'Lifestyles which combine Health And Sustainability'. They estimate 50 million Americans follow or aspire to such a lifestyle, and these are no 'new age fantasists', but the consumers who have created a market for products and services estimated at US$230 billion in the US and US$540 billion globally. Food for health stands within this framework and is inevitably influenced by the changes consumers are making in other areas of their lives.

Turning away from past efforts like boycotts, refusing to invest in certain companies or putting resolutions to company boards, socially conscious investors are now putting their money where their mouths are. For example, a report published in 1999 by the Social Investment Forum, a non-profit trade organization, estimates that more than US$2 trillion is invested along socially conscious lines in the US, accounting for about 13 per cent of all money under professional management. This was up by 82 per cent from 1997. At the same time, the report says, the number of socially responsible mutual funds rose to 175 from just 55 in 1995 (reported in *Financial Times Food Business*, February 2000).

What are the implications of these trends for the marketing of functional foods? Put very simply, companies who want to enjoy long-term success should follow both the magnetic pull of consumer trends and the shift in public health thinking and take the 'decency' approach to food and health as the basis for their thinking and product development. Even though a product may address a single risk factor it will still have to be marketed within the holistic framework of health.

A Healthful Company™ builds relationships with consumers

Popcorn (1991) said that the future would see the fragmentation of mass markets and the gradual customization of products to meet the needs of smaller groups of consumers. This is now making itself felt in the food industry, with the idea of individualized nutrition, even customized diets, gradually gaining wider currency. The proliferation of wheat-free, gluten-free and other diets shows that consumers are experimenting with ways to improve their health through diet on a highly individual basis. Companies will have to respond to this – there will be no point treating these consumers as dietary 'cranks' – because on current trends the day is not far away when we will all be cranks.

The Healthful Company™ first steps back and tries to engage in this more personal or emotional relationship we have with health, for example by looking at health – and you will note we talk about health not nutrition – in the broader context. It is worth remembering that health is also 'big' in many other industries, from cosmetics to country clubs.

Popcorn is alluding to what is more formally treated in the business literature as 'relationship' marketing. Brand and image building, marketing communications, are key components of building relationships with customers (Achrol and Kotler, 1999), but many companies struggle to achieve truly successful relationships with their consumers. For The Healthful Company™ traditional marketing skills are no less important, but the company's approach to them and how they go about them is informed by a holistic approach to health and its desire to be at the forefront of relationship marketing.

Companies are still being urged to appreciate that their most valuable assets are relationships with customers: as Duncan and Moriarty (1993, p3) say:

> *'Ongoing customer relationships are the company's most important business asset. Estimates that it costs six to nine times more to acquire a new customer than it does to retain a current one demonstrate the value of relationships. In addition, profits per customer increase with customer longevity, because the longer customers are with a company, the more willing they are to pay premium prices, make referrals...and spend money. The more a company can do to strengthen customer and other stakeholder relationships, the more cost-effective its marketing effort will be.'*

Relationship marketing is particularly important for functional food because of the higher level of investment needed than for traditional food and beverage research and development. With higher up-front costs in developing functional foods, long-term rewards become more important, as does the development of a sustainable model of business as a source of competitive advantage. The model of The Healthful Company™ meets these challenges.

Creating strong trusting relationships and emotional attachments with consumers has become a massive problem for the food industry. Much of the food industry has lost both credibility and the confidence of many consumers. Consumers are much more sceptical about 'expert' reassurances as a result of hearing conflicting messages in the media about food and health, and in the wake of the BSE crisis and other food safety scares. Using the example of modern medicine,

Williams and Calnan (1996) write: 'Trust has to be increasingly won and maintained in the face of growing public criticism and scepticism.' Their comments can apply equally to the functional foods revolution.

The Healthful Company™ makes an offer that is both product and service

Here, again, we use a popular business book to illustrate our theme. Stan Davis and Christopher Meyer in their book *Blur* summarize many of the themes of the new economy. For these authors the new economy is about the way people will use resources to fulfill their desires and is driven by what they call 'connectivity, speed, and the growth of intangible value' (Davis and Meyer, 1998). An important aspect of the new economy for these authors is that the historic distinction between product and service will become less relevant. Increasingly, businesses will have to make consumers an 'offer'. They say 'winners provide an offer that is both product and service simultaneously'.

People's health and well-being is more than dependent on product alone. The Healthful Company™ willl work out ways of delivering a health service as much as a product with a health benefit. The challenge is to develop and deliver the service together with the product. There is nothing new in this idea and as we point out it has become common parlance even in the popular business literature in recent years. What would be new is for food companies to work out how to become sufficiently innovative and creative to do this successfully for functional food. Like the US drug companies changing themselves to become disease management organizations, thus blurring service and product, could functional food companies aspire to become health management businesses?

Telling a great story – The Healthful Company™ is a great communicator

We are big fans of Rolf Jensen, director of the Copenhagen Institute for Future Studies, and as a result believe The Healthful Company™ should see itself as a full participant in what he describes as the Dream Society (Jensen, 1999). Jensen says that businesses need to imagine their futures the way good novelists imagine their stories. The future of business won't be the latest technology or newest product, but the story behind the product that will provide the competitive edge. With the unique role food plays in our lives – its deep cultural and social meanings; its biological significance; its meaning for health and well-being – all Healthful Companies should be Dream Society companies. As Jensen says:

> *'The Dream Society is emerging this very instant – the shape of the future is visible today. Right now is the time for decisions...now is the time to add emotional value to products and services.'*

A compelling aspect of the Dream Society is the ability of companies to tell a 'great story'. Jensen says that in the Dream Society, our work will be driven by stories and emotions, not just data. The story – or 'value statements' as he defines a story – will become an ever-more important part of the decision to buy.

Jensen gives a food-related example of how the Dream Society logic is being used today. He points out that in Denmark eggs from free-range hens have captured over 50 per cent of the market. He says consumers do not want hens to live their lives in small, confining cages; they want hens to have access to earth and sky. Consumers desire eggs to be produced under the technology and methods of our grandparents – the old-fashioned way. (Incidentally, they would most likely become 'functional' eggs, having a better essential fatty acid profile than modern mass-produced battery eggs.) This means the eggs become more expensive – more labour intensive – but consumers are happy to pay an additional 15 to 20 per cent for the 'story behind the eggs'. Jensen says they are willing to pay more for the story about animal ethics, rustic romanticism and the good old days. He points out that in five to ten years this could be the way eggs will be produced everywhere.

We are not suggesting everything should become free-range, but this example from Jensen illustrates the power of the Dream Society. With a story about hens and eggs a whole industry and market is being transformed – the old industry possibly becoming outlawed – and the egg market has acquired a new dimension: it no longer consists of a standardized commodity, mass-produced at the lowest price.

Thinking the unthinkable will become commonplace for business survival and success. The Healthful Company™ knows how to tell and practice a 'great story' about itself, food and health.

The Healthful Company™ is knowledge-rich

To achieve the five points outlined above The Healthful Company™ has to know a lot, know how to use this knowledge and to apply the relevant parts. Part of The Healthful Company™'s unique knowledge base is its science and technology – its know-how which is the rock on which it rests.

Knowledge is also the key to a company's communications to consumers and the media. It is the cornerstone of credibility. While many scientists are often brilliant at explaining functional food science face-to-face, and the nuances and the bigger picture behind their work, this clarity becomes lost along the food communications chain. The skill of communication is the key, knowledge is power, and firms that use it tend to go further ahead than their competitors. Investing in know-how and being able to leverage this will see certain functional food companies pull ahead of the rest.

The Healthful Company™: Can the New Business Model for Food and Health be Achieved in Practice?

Earlier we outlined the characteristics of the model of The Healthful Company™. Can they be achieved in practice? We have looked at many companies and have come to admire greatly the work of a small number of them. For example, the Belgium based company Orafti has almost single-handedly championed the value of prebiotics throughout the world (see Chapter 9) and in doing so has made the world's regulators and policy makers reconsider the whole issue of dietary fibre and human health. The Finnish dairy company Valio has championed the science of probiotics (see Chapters 8 and 9). In both these cases there have so far been limits to their achievements in terms of becoming fully fledged Healthful Companies. Orafti, as a food ingredient company, has no presence in the marketplace and therefore is one step removed from being able to establish its relationship directly with consumers. Valio has taken the route of licensing its know-how to third parties so, except in its home country Finland, plays the role of technical support and is again one step removed from establishing a healthful relationship with consumers.

We can only think of one company that comes close to actively being a Healthful Company™ in the functional foods revolution on the international stage, and that is the Japanese-based company Yakult Honsha. In fact, through studying Yakult, the company became our inspiration for the concept of The Healthful Company™.

Yakult was founded with the health and well-being of the consumer at its core. It is this philosophy that informs its business to this day. As we described in Chapter 6, the founder of Yakult, Dr Minora Shirota had a simple but powerful message about health:

prevent disease rather than treat disease; a healthy intestine leads to a long life; and, deliver health benefits to as many people as possible at an affordable price.

The idea behind The Healthful Company™ is about companies behaving like their consumers, putting health at the centre of corporate life just as their consumers put it at the centre of their lives. The Yakult philosophy is an example of the power of the holistic approach to health. The philosophy of the company's founder Dr Minora Shirota and the constant retelling of his story reinforces this approach. It is no accident that the Yakult company motto is 'Working on a healthy society'.

Yakult has also demonstrated that The Healthful Company™ can play hard in the marketplace. For example, it has successfully created an entirely new and innovative category in Europe, forcing Europe's (and some of the world's) largest dairy companies into becoming market-followers and to rethink many aspects of their strategies for functional foods. This is a particular achievement since Yakult is not itself a dairy company but a company focused on products for human health based on probiotic lactic acid bacteria.

Yakult has created a unique consumer franchise and relationship in Europe and other countries. This has meant talking to consumers about the health of their gut, introducing ideas like 'friendly bacteria' and helping consumers to understand the daily-dose idea behind its little-bottle concept.

In contrast, the results which Yakult's emulators achieve are often, at best, lacklustre. Nestlé, for example has an entire division devoted to nutrition and research and development far bigger than Yakult's. Europe is Nestlé's home market, its brand recognition is probably among the highest in the world – while Yakult was unheard of in Europe five years ago – and yet Nestlé's LC1 Go!, its answer to Yakult, has achieved a market share of just 5 per cent in the UK, compared to Yakult's 75 per cent. The reason for this difference in performance, we believe, arises from Yakult's unique approach to doing business – an approach that is the foundation of our Healthful Company™ model.

Health at the Heart of the Corporate Soul

Yakult's philosophy recognizes that healthfulness is based on more than a product – it is about a service. In many countries the service is literally delivered face-to-face through Yakult 'ladies'. In other countries extensive face-to-face promotional activities are employed. Yakult in effect supplies a 'health service' through, in this case, a product that supports health. The company also goes to extraordinary lengths to provide an education service to inform consumers about how to maintain general health through digestive health through, for example, publicatons such as the *Yakult Guide to the Gut* – which provides frank information about stools and the working of the intestines.

Yakult lives its philosophy 'Working on a healthy society' by providing a wider service through its sponsorship of activities that the company describes as contributing to the health of society at large. An example from the UK is the company's little-publicized provision of sponsorship for art students from Britain's Royal Academy to visit schools in deprived areas to bring to children their art teaching. These schoolchildren from disadvantaged backgrounds are not the upscale consumers usually courted by many companies active in the functional foods revolution. This is true philanthropy.

Yakult conducts all its public relations in-house. So important is the business of communicating knowledge that the company believes that it can only be entrusted to the company's own employees – people who believe in, are committed to, and understand the company's philosophy and put into practice the concept of working for a healthy society. The Yakult story is a 'great story'. We would love to hear stories from other companies who are, or are on their way to becoming, Healthful Companies.

Conclusion

We have set out the challenges and opportunities of the functional foods revolution for public health professionals, policy makers, nutritionists and corporations. For corporations we have presented The Healthful Company™ as a model for sustainable and profitable success in the functional foods revolution. Whether companies do decide to adopt the model of The Healthful Company™ is up to their managements. Whether they think they can is best answered by Henry Ford, who said:

> *'Whether you believe you can, or whether you believe you can't, you're absolutely right.'*

And if you want to find out how the functional foods revolution unfolds, and which companies have the courage and skill to become truly Healthful Companies, you can keep track of developments at our website (www.new-nutrition.com).

Finally we, no less than The Healthful Company™, believe in the importance of open relationships and good communications with our customers – in this case you the reader. You have heard from us, now we look forward to hearing from you. To remind you, we can be contacted at: michaelh@nutritiondigest.com (Michael Heasman) and julian@nutritiondigest.com (Julian Mellentin).

Appendix 1

Summary of the UK Code of Practice on Health Claims on Food

General Principles for Making a Health Claim

The Code sets out a series of principles that should be followed when judging whether a health claim is misleading, false or exaggerated. The principles in the Code are applied to the health claim as it appears to the consumer as well as its literal or legal meaning. For example, marketing imagery, careful use of words and literature given out with the product are all taken into account.

The principles

- Health claims must be truthful, consistent with the evidence and must not mislead, exaggerate or deceive. It must be clear who will benefit, for example, everyone, the elderly, women, children etc;
- A health claim must not encourage over-consumption or be negative about other normal or healthy foods or the need for a good overall diet. Health claims must be put into the context of the overall diet;
- The benefit claimed must generally come from the single food for which the claim is made when eaten in the quantities reasonably expected to be consumed in one day. However, health claims can also be made about the value of different foods working together;
- Health claims should not be made for foods that are unlikely to provide a significant contribution to a healthy diet (eg 'low salt candy-floss is good for the heart');
- Medical claims cannot be made (ie that the product can treat a disease) because this is illegal. However, it is acceptable to refer to maintaining health or to risk factors which affect health. The enforcement authorities in the field have undertaken to help

companies if a health claim needs to be explicit about disease in order to communicate effectively to the consumer. The Code preamble also calls for a review in the law to allow the connection between disease and diet to be more clearly communicated;

- If health claims are made, the Code also suggests that certain information about the product other than the minimum required by law is also given to consumers. However, this is not required to comply with the Code.

The other main areas of the Code of Practice deal with:

- Substantiation and pre-market advice;
- Sources of evidence;
- Documentation required;
- The Code Administration Body;
- Suspected breaches of the Code;
- Regular review.

Three annexes are included:

1 Annex I. Sets out detailed guidance on 'borderline' claims. Advice is also given on how to communicate health claims without breaking the law and on the procedure to follow where an explicit reference to a disease may be needed in order to effectively communicate a health claim to the consumer;
2 Annex II. Provides advice on setting up trials in support of health claims;
3 Annex III. Sets out the options for the Expert Authority and the basic requirements to ensure its credibility.

Appendix 2

The Definition of FOSHU

Foods covered by the definition of 'foods for specified health use' (FOSHU) include those consumed in the normal way and exclude drugs in tablet or capsule form. These foods should support the improvement of the diet and help maintain or improve health. The labelling of such foods should provide nutritionally and/or medically correct information on the relationship between this food or ingredient and health. In particular, three conditions that define a functional food are highlighted:

1. It is a food (not a capsule, tablet or powder) derived from naturally occurring ingredients;
2. It can and should be consumed as part of the daily diet;
3. It has a particular function when ingested, serving to regulate a particular body process, such as:
   - enhancement of the biological defence mechanism;
   - prevention of a specific disease;
   - recovery from a specific disease;
   - control of physical and mental conditions; and
   - slowing the ageing process.

The Japanese Ministry of Health has defined FOSHU which:

- are expected to have a specific effect on health due to relevant constituent(s) of the foods, or
- are foods from which allergens have been removed, and
- the effect of such addition or removal has been scientifically evaluated, and
- permission has been granted to make claims regarding the specific beneficial effect on health to be expected from their consumption.

FOSHU foods should promote health by supporting the improvement of the diet. The food should not pose a health or hygiene risk.

The labelling of FOSHU should provide the following information:

- The designation of the food (and trade name);
- The date of manufacture;
- The name and address of the manufacturer and the distributor;
- The net contents (weight or volume);
- The reason for approval (effect on health) and general guidelines on health;
- A table of nutritional values and calories;
- The names of the ingredients (listed in descending order);
- The date of minimum durability (the period within which the quality may be maintained under specified storage conditions after manufacture of the food);
- Indication that this is a 'food for specific health use', the recommended intake and a warning against excessive intake;
- Any special cautions relating to intake, cooking or storage where necessary, under the heading 'Cautions';
- The name and address of the person to whom the approval was granted where this person is not the manufacturer.

In addition, the Japan Health Food and Nutrition Association recommends that the following information should be supplied to consumers of FOSHU:

- It is important to maintain a balanced diet;
- The FOSHU is expected to be good for your health, therefore it should be consumed as part of the daily diet, replacing ordinary foods;
- Pay attention to the directions for use, such as the anticipated effects and the recommended intake;
- The labelling has been approved by the Ministry of Health.

Appendix 3

Main Provisions of the Dietary Supplement Health and Education Act of 1994 (DSHEA)

Definition of Dietary Supplement

> *'A product...intended to supplement the diet that bears or contains...a vitamin;...mineral; herb or other botanical;...an amino acid;...a dietary substance for use by man to supplement the diet by increasing the total dietary intake; or...a concentrate, metabolite, constituent, extract, or combination (of any ingredients) described above.'*

There are additional criteria. A dietary supplement must, either be intended for ingestion in tablet, capsule, powder, soft gel, gelcap or liquid droplet form, or, if not intended for ingestion in such a form, must not be 'represented for use as a conventional food or as a sole item of a meal or the diet'. Products must be labelled 'dietary supplement'.

Statements of Nutritional Support

The DSHEA establishes exceptions that allow dietary supplements to make four types of 'statements of nutritional support' on labels or in other labelling without obtaining FDA approval of a health claim regulation. These exceptions include:

- A statement that 'claims a benefit related to a classical nutrient deficiency disease and discloses the prevalence of such disease in the United States';
- A statement that 'describes the role of a nutrient or dietary ingredient intended to affect the structure or function in humans';

- A statement that 'characterizes the documented mechanism by which a nutrient or dietary ingredient acts to maintain such structure or function';
- A statement that 'describes general well-being from consumption of a nutrient or dietary ingredient'.

Any of the above types of statements may be made in labelling for a dietary supplement, without the approval of a health claim regulation, if:

- The manufacturer 'has substantiation that such statement is truthful and not misleading';
- The labelling contains, prominently displayed, the following additional text: 'This statement has not been evaluated by the Food and Drug Administration. This product is not intended to diagnose, treat, cure, or prevent any disease';
- The manufacturer notifies the FDA 'no later than 30 days after the first marketing of the dietary supplement with notification that such a statement is being made'.

In addition, the DSHEA greatly restricts the FDA's previous ability to object to the use of books and other publications in connection with the sale of dietary supplement products. The legislation provides that 'a publication, including an article, a chapter in a book, or an official abstract of a peer-reviewed scientific publication that appears in an article and was prepared by the author or the editors of the publication'... shall not be defined as labelling' and may 'be used in connection with the sale of a dietary supplement to consumers' if the publication is 'reprinted in its entirety' and meets certain specific criteria (eg it must not 'promote a particular manufacturer or brand of a dietary supplement').

Notes

Chapter 1

1 The first European consultancy report on functional foods was by PA Consulting. Reading it again, ten years later, the report can be regarded as a rarity for its accurate insight into future developments. It makes the point that in 1989 few major European companies had even heard of functional foods, but suggests that profit margins would be substantial – in the order of 40 per cent (PA Consulting, 1990) which for some subsequent functional products has proved a gross understatement!

2 The chronology in this box is not meant to be exhaustive, but indicative of developments. There is a bias in the selection of landmark events with an emphasis on developments in the US and the UK and even here we have been selective. The box is, therefore, not meant to imply that food and health initiatives in other countries are not also ground-breaking or to lessen the importance of these. In fact it would be possible to develop similar detailed chronologies for many countries.

3 It should also be noted that the term modified in this definition does not refer to genetically modified foods or ingredients, but is used in the normal meaning of the word.

4 For a detailed market overview see the management report by Heasman and Mellentin (1998).

5 Folates represent an important group among the B-vitamins. Folate is a generic term for the naturally occurring forms of B-vitamin, whereas folic acid refers to the synthetic compound most often used in vitamin supplements and in fortified products.

6 Put simply, Fordist thinking, or Fordism, describes a society in which a system of mass production and consumption is the dominant system of producing goods. It is derived, as the name implies, from principles first developed by Henry Ford in making the Model T Ford motor car using, for the first time, the automated assembly line.

7 Market figures for functional foods vary considerably depending on definitions used, the reliability of the market analysts and the product categories reviewed. They should all be treated with caution.

Chapter 2

1 Plant sterols are found in relatively small quantities in nature, but their potential effects on lowering cholesterol had been reported in the literature since the 1950s. However, they are not readily available and are difficult to use in product development.
2 McNeil works closely with Raisio, who keep control of the production of plant stanol esters used for the products McNeil sells.
3 Unilever say that this is not genetically modified – it is GM-free.

Chapter 4

1 One of the major concepts of the modern world is the idea of risk and strategies and actions to manage risk. Risk, like that other all-encompassing, and related, concept of recent years, globalization, is coming to symbolize the way in which we view the world in the 21st century. In particular, in the sociology literature a number of theoretical perspectives on risk (Lupton, 1999) have emerged, but it is the work in one of these areas, as developed by Anthony Giddens and Ulrich Beck, particularly Beck's concept of the risk society, that is probably best known in popular discussions (see for example Beck, 1992; Giddens, 1998). The idea of the risk society is an attractive one, especially when considering the global food economy. Put simply, in Beck's view, Western societies are living in a transitional period, in which industrial society is becoming risk society. By this, again putting it very simply, Beck means risks have proliferated as an outcome of modernization and we are seeing debates and conflicts over risks begin to dominate public, political and private arenas. Beck says individuals are now forced to deal with risks on an everyday basis: 'Everyone is caught up in defensive battles of various types, anticipating the hostile substances in one's manner of living and eating.' (Beck, 1994)

2 In a similar vein, Williams and Calnan (1996) point out when confronted by modern medicine, individuals are not simply passive consumers who are duped by medical ideology. Rather they are reflexive agents who are active in the face of modern medicine and technological developments.

3 In contrast to the medicine, probabilistic thinking has already triumphed in the natural and social sciences, in policy for education, social insurance, public health and business, for example, through quality improvements (Fox 1999). Goodman (1999) suggests some historical reasons for this. Little to do with science, more to do with power and status. In the 18th century, for example, the use of 'aggregate thinking' was denounced, in this case by the French Academy of Medicine, as distracting from the physician's primary role of understanding and properly treating the individual patient. As Goodman writes: 'The medical profession long resisted the notion that frequencies of past events informed us about future ones, one reason being that this idea relegated clinical explanation to a secondary role. It undermined the authority of individual physicians by threatening the proposition that the uniqueness of each patient demanded uniquely tailored care.' (p. 604) In other words, the debate about the meaning of an individual patient's risk is centuries old. The more recent discussion over the patient's relative risk and a population's absolute risk (the 'number needed to treat') is a modern manifestation of this 18th-century debate. If anything, functional food science threatens to sharpen this debate.

Chapter 5

1 Directive 89/398 on Foods for Particular Nutritional Uses ((PARNUTS) is designed to control the composition and labelling of infant formulae and baby foods, and of foods which have been formulated for:

- people whose digestive processes or metabolism are disturbed;
- people who are in a special physiological condition and who are therefore able to obtain special benefit from controlled consumption of certain substances.

The categories of foods listed in the Directive as coming within this definition are slimming foods, low-sodium foods, gluten-free foods, sportsmen's foods, diabetic foods and medicinal foods.

2 It is only right we declare an interest here. Author Heasman, although not on the functional foods working party, was a member of the National Food Alliance's working party on Food Labeling and Advertising at the time and was involved in early meetings specifically on functional foods. Heasman remains a member of Sustain's (as the NFA has been renamed) working party on Food Labeling and Advertising, but did not take part in any of the meetings of the Joint Health Claims Initiative.

3 The NCC was set up by the UK government in 1975 to represent the interests of UK consumers of goods and services of all kinds, both in the public and private sectors. It is an independent body which campaigns, conducts research and supports other consumer organizations. It has a particular responsibility to represent disadvantaged consumers.

4 Red Bull, an energy drink, has become one of the most successful new product introductions in the UK. Through brilliant lifestyle marketing, starting with the UK's club culture, it is now widely available in Britain's supermarkets, where it is predicted in 2000 to start outselling colas. Although Red Bull is not attempting to make a public health contribution, and hence we exclude it from detailed analysis in this book, its marketing warrants very close study by producers of functional food on how to build marketing relationships with consumers, introduce an unusual product concept and target it expertly in conjunction with lifestyle expectations. Another unusual point to note in its phenomenal success in the UK (and other European countries), is that to both the middle-aged authors, Red Bull tastes like yuk!

Chapter 6

1 See chapter 9 for a detailed description of probiotic lactic acid bacteria.

2 There are very few sources on the Japanese food industry written in English. This chapter relies on three main sources, *Japanscan Food Industry Bulletin*; information from the Japan Health Food & Nutrition Food Association; and from interviews undertaken by the authors. We would particularly single out *Japanscan Food Industry Bulletin* as an outstanding source of information on the Japanese food industry in general. Japanscan is published monthly as a newsletter and can be contacted at Anville, Upper

Quinton, Stratford-on-Avon, CV37 8SX, United Kingdom. Tel: +44 (0)1789 720 395; fax: +44 (0)1789 721 808; e-mail: japanscan@compuserve.com. Japanscan also publish market reports on different aspects of the food industry in Japan and there has been a regular series on functional foods. Much of the company information in this chapter relies on Heasman and Mellentin (1998) which used material specially commisioned by the authors from Japanscan. At the Japan Health Food & Nutrition Food Association we would especially like to thank Kaori Nakajima, responsible for Scientific and Regulatory Affairs, for her help in providing information. We also acknowledge Yakult UK who facilitated a study trip to Japan for us in 1999 and the Japanese experts who gave their time during this trip. But in all cases, it should be noted the interpretation and analysis presented here, and any errors, are the responsibility of the authors.

Chapter 7

1 In the US, ingested products that have health-related benefits to which reference is made in labeling generally will fall into either or both of two regulatory categories: 'food' and/or 'drug'. 'Food' means 'articles used for food or drink for man or animals', while 'drug' means 'articles intended for use in the diagnosis, cure, mitigation, treatment, or prevention of disease' or 'articles (other than food) intended to affect the structure or any function of the body'. In addition, in the US a 'health claim' is legally defined. In general it includes: 'any claim made on the label or in [other] labeling of a food, including a dietary supplement, that expressly or by implication...characterizes the relationship of any substance to a disease or health-related condition'. 'Disease or health-related condition' is defined to mean: 'damage to an organ, part, structure, or system of the body such that it does not function properly (eg cardiovascular disease), or a state of health leading to such dysfunctioning (eg, hypertension); except that diseases resulting from essential nutrient deficiencies (eg scurvy, pellagra) are not included....'

2 For the main provisions of the Dietary Supplement Health and Education Act of 1994, see Appendix Three.

Chapter 9

1 Although in this chapter we focus on functional foods in the dairy industry based on gut health and pro and prebiotics, the dairy industry has also been active in other areas of health promoting foods. For example, Parmalat, Italy's largest dairy company developed and launched a milk product with omega-3 fatty acids called "Plus Omega 3" and packaged and promoted with a heart logo as "the milk of life". Other dairy companies have also introduced spreads and/or margerines enriched with omega-3, also for heart health. We have chosen to focus on gut health and probiotic products as one of the most dynamic and extensive of functional food markets in Europe and one with a clear international focus.

References

The American Dietetic Association (1995) Position of the American Dietetic Association: Phytochemicals and Functional Foods, *Journal of the American Dietetic Association*, vol 95, no 4, pp493–496

Achrol, R and Kotler, P (1999) Marketing in the Network Economy, *Journal of Marketing*, vol 63 (Special Issue 1999), pp146–163

Advertising Age, December 1999

Affertsholt, T (2000), FoodGroup, Denmark, personal communication

The Alpha-tocopherol, Beta-carotene Cancer Prevention Study Group (1994) The effect of vitamin E and beta carotene on the incidence of lung cancer and other cancers in male smokers, *The New England Journal of Medicine*, vol 330, no 5, pp1029–1035

Arai, S (1996) Studies on functional foods in Japan – state of the art, *Biosciences, Biotechnology, Biochemistry*, vol 60, no 1 pp9–15

Avorn, J, Monane, M, Gurwitz, J, Glynn, R, Choodnovskiy, I, and Lipsitz, L (1994) Reduction of bacteriuria and pyuria after ingestion of cranberry juice, *JAMA*, vol 271, no 10, pp751–754

Balanya, B, Doherty, A, Hoedeman, O, Ma'anit, A, and Wesselius, E (2000) *Europe Inc,* London, Pluto Press

Beck, U (1992) *Risk Society: Towards a New Modernity*, London: Sage

Beck, U (1994) The reinvention of politics: towards a theory of reflexive modernization, in Beck, U, Giddens, A and Lash, S, *Reflexive Modernization: Politics, Tradition and Aesthetics in the Modern Social Order*, Cambridge: Polity Press, pp1–55

Bellisle, F, Diplock, A, Hornstra, G, Koletzko, B, Roberfroid, M, Salminen, S, and Saris, W (1998) Functional Food Science in Europe, *British Journal of Nutrition*, vol 80, (suppl 1), S1–S193

Bengoa, JM (1997) A Half-Century Perspective on World Nutrition and the International Nutrition Agencies, *Nutrition Reviews*, vol 55, no 8, pp309–314

Bloom, BR (1999)The Future of Public Health, *Nature,* vol 402, suppl C63–C64

Boon, T (1997) Agreement and Disagreement in the Making of 'World Of Plenty', pp166–189 in Smith (1997)

Bradbury, J, Lobstein, T, and Lund, V (1996) *Functional Food Examined. The Health Claims Being Made for Food Products and the Need for Regulation*, London, The Food Commission

Business Week, (1989) Can Cornflakes Cure Cancer?, 9 October, cover

Breslow, L (1999) From disease prevention to health promotion, *JAMA*, vol 281 no 11 pp1030–1033

Bruce, S and Crawford, B (1995) *Cerealizing America*, London, Faber and Faber

Bush, L and Williams, R (1999) Diet and health: new problems/new solutions, *Food Policy*, vol 24, pp135–144
Callaway, CW (1997) Dietary Guidelines for Americans: An Historical Perspective, *Journal of the American College of Nutrition*, vol 16, no 6, pp510–516
Cannon, G (1987) *The Politics of Food*, London, Century Hutchinson Ltd
Cannon, G (1988) Diet and the Food Industry, *RSA Journal*, vol CXXXVI, no 5382, May, pp 398–416
Cannon, G (1992) *Food and Health: The Experts Agree*, London, Consumers' Association
Cannon, G (1995) The new public health, *British Food Journal*, vol 95, no 5, pp4–11
Caplan, P (ed) (1997) *Food, Health and Identity*, London, Routledge
Cockbill, C (1993) Food Law and Functional Foods, *British Food Journal*, vol 96, no 3, pp3–4
Colles, L(1998) *Fat: Exploding the Myths*, London, Carlton Books
COMA (Committee on Medical Aspects of Food Policy) (1984) *Diet and Cardiovascular Disease*, Report on the panel on Diet in Relation to Cardiovascular Disease, London, HMSO
Conning, D (1995) *A New Diet of Reason*, London, The Social Affairs Unit
Consumers' Association (2000) *Functional foods – health or hype?*, London, Consumers' Association
Coussement, P (1997) Powerful Products, *The World of Ingredients*, August, pp12–17
Cowley, G (2000)How to Get to Your Golden Years, *Newsweek*, 3 April, pp72–74
Clinton, S (1998) Lycopene: Chemistry, Biology and Implications for Human Health and Disease, *Nutrition Reviews*, vol 56, no 2, pp35–51
Clydesdale, FM and Chan Soh Ha (eds) (1996) First International Conference on East-West Perspectives on Functional Foods, Singapore, 26–29 September, 1995, proceedings published in *Nutrition Reviews*, vol 54, no 11 (part II), ppS1–S202
Clydesdale, FM (1997) A Proposal for the Establishment of Scientific Criteria for Health Claims for Functional Foods, *Nutrition Reviews*, vol 55, no 12, pp413–422
Cwiertka, K (1998) A note on the making of culinary tradition – an example of modern Japan, *Appetite*, vol 30, pp117–128
Cwiertka, K (1999) Culinary globalization and Japan, *Japan Echo*, vol 26, no 3 pp52–58
Davis, S and Meyer, C (1998) *Blur: The Speed of Change in the Connected Economy*, New York, Warner Books
Dawber, T (1980) *The Framingham Study: The Epidemiology of Atherosclerotic Disease*, Cambridge, MA, Harvard University Press
DeFelice, S (1995) *Cereal Foods World*, vol 40, pp51–52 or The nutraceutical revolution: its impact on food industry R&D, *Trends in Food Science and Technology*, vol 6, pp59–61

DeFelice, S (1998) *Nutraceuticals. Developing, Claiming, and Marketing Medical Foods*, New York, Marcel Dekker Inc

Dietary Goals for the United States (1977) Select Committee on Nutrition and Human Needs, United States Senate, Washington, DC, US Government Printing Office

Diplock, A, Aggett, P, Ashwell, M, Bornet, F, Fern, E, and Roberfroid, M (1999) Scientific concepts of functional foods in Europe: Consensus document, *British Journal of Nutrition*, vol 81, supplement no 1, ppS1–S27

Dragon (1999) *New Foods: The Future of Positive Health*, London, Dragon International

Duncan and Moriarty (1998) A communications based marketing model for managing relationships, *Journal of Marketing*, April, vol 62, pp1–13

Ernst & Young and A C Nielsen (1999) *New Product Introduction, Successful Innovation/Failure:* Fragile Boundary

ESRC Global Environmental Change Programme (1999) *The politics of GM food: risk, science & public trust*, Special Briefing No 5, available from www.gecko.ac.uk/gm-briefing.html

Euromonitor (2000) *Functional Foods: A World Survey*, London, Euromonitor

Fast Company, March 2000

FIND/SVP (1994) *The Market for Nutraceutical Foods and Beverages*, New York

Finnish Food and Drink Industry's Federation (1999) *Statistical Review*, Helsinki

Fischer, A (1998) *Is Your Career Killing You?* Data Communications, February

Forde O (1998) Is Imposing Risk Awareness Cultural Imperialism? *Social Science and Medicine*, vol 47, no 9, pp1155–1159

Fox, D (1999) Comment: Epidemiology and the New Political Economy of Medicine, *American Journal of Public Health*, vol 89, no 4, pp493–496

Fuller, R (1991) Probiotics in Human Medicine, *Gut*, vol 32, pp439–442

Fuller, R (ed) (1992) *Probiotics. The Scientific Basis*, London, Chapman & Hall

(FFH) Functional Foods for Health Program (1999): *Functional Foods for Health News*, vol 6, no 4, University of Illinois at Chicago and Urbana-Champaign

Germov, J and Williams, L (eds) (1999) *A Sociology of Food and Nutrition*, Oxford, Oxford University Press

Gerster, H (1997) The Potential Role of Lycopene for Human Health, *Journal of the American College of Nutrition*, vol 16, no 2, pp109–126

Gibbons, M (1999) Science's New Social Contract with Society, *Nature*, vol 402 suppl, ppC81–C84

Gibney, M and Strain, S (2000) Food and nutrition for all, *The Lancet*, vol 354, psiv38

Gibson, G and Roberfroid, M (1995) Dietary Modulation of Human Colon Microbiota: Introducing the Concept of Prebiotics, *Journal of Nutrition*, vol 125, pp1401–1412

Giddens, A (1998) Risk Society: the context of British Politics, in Franklin, J (ed) *The Politics of Risk Society*, Cambridge, Polity Press, pp23–34

Giovannucci, E, Ascherio, A, and Rimm, E (1995) Intake of Carotenoids and Retinol in Relationship to Risk of Prostate Cancer, *Journal of the National Cancer Institute*, vol 87, pp1767–1776

Goldberg, I (ed) (1994) *Functional Foods, Designer Foods, Pharmafoods, Nutraceuticals*, London, Chapman & Hall

Goldin, B (1998) Health Benefits of Probiotics, *British Journal of Nutrition*, vol 80, suppl 2, ppS203–S207

Goodman, D and Redclift, M (1991) *Refashioning Nature: Food, Ecology and Culture*, London and New York, Routledge

Goodman, S (1999) Probability at the Bedside: The Knowing of Chances or the Chances of Knowing, *Annals of Internal Medicine*, vol 130, no 7 pp604–606

Hanson, L and Yolken, R (eds) (1999) Probiotics, Other Nutritional Factors, and Intestinal Microflora, *Nestle Nutrition Workshop Series*, vol 42, Switzerland, Nestec Ltd

Hasler, C (1996) Functional Foods: The Western Perspective, *Nutrition Reviews*, vol 54, no 11 ppS6–S10

Hasler, C (1998) A new look at an ancient concept, *Chemistry & Industry*, 2 February, pp84–89

Hasler, C (1998) *Functional Foods: Institute of Food Technologists Scientific Status Paper*, Chicago, IFT

Hasler, C, Huston, RL and Caudill, EM (1995) The Impact of the Nutrition Labeling and Education Act on Functional Foods, in Shapiro, (1995)

Hawken, P, Lovins, A and Lovins, L (1999) *Natural Capitalism*, New York, Little, Brown & Company

Health Education Authority (1997) *Eight Guidelines for Healthy Diet: A Guide for Nutrition Educators*, Abingdon, Health Education Authority

Heasman, M (1999a) *The Functional Foods Revolution. A New Nutrition Agenda for a New Century*, The Caroline Walker Lecture 1999, London, The Caroline Walker Trust

Heasman, M (1999b) A healthy Vision from Japan, *New Nutrition Business*, vol 4, no 8, p20

Heasman, M and Fimreite, D (1998) Functional Foods: Breaking Boundaries, *Nutrition Science News*, vol 3, no 7, pp366–372

Heasman, M and Mellentin, J (1998) *The Business of Healthy Eating: Global Trends, Developments and Strategies in Functional Foods and Nutraceuticals*, London, Financial Times Retail & Consumer Publishing

Heasman, M and Mellentin, J (1999) Functional Foods are Dead, Long Live Functional Foods, *New Nutrition Business*, vol 4, no 7, pp16–19

Heimburger, D, Stallings, V, and Routzahn, L (1998) Survey of Clinical Nutrition Training Programs For Physicians, *American Journal of Clinical Nutrition*, vol 68, no 6, pp1174–1179
Helsing, E (1989) Nutrition Policies in Europe – the State of the Art, *European Journal of Clinical Nutrition*, vol 43, suppl 2, pp57–66
Helsing, E (1997) The History of Nutrition Policy, *Nutrition Reviews,* vol 55, no 11, ppS1–S3
Hennekens C, Buring, J, Manson, J, Stampfer, M, Rosner, B, Cook, N, Belanger, C, LaMotte, F, Gaziano, J, Ridker, P, Willett, W, and Peto, R (1996) Lack of effect of long-term supplementation with beta carotene on the incidence of malignant neoplasms and cardiovascular disease, *New England Journal of Medicine*, vol 334, pp1145–1149
Holzapfel, W, Haberer, P, Snel, J, Schillinger, U, and Huis in't Veld, J, (1998) Overview of Gut Flora and Probiotics, *International Journal of Food Microbiology*, vol 41, pp85–101
Hornstra, G, Barth, C., Galli, C, Mensink, R, Mutanen, M, Riemersma, R, Roberfroid, M, Salminen, K, Vansant, G, and Verschuren, P (1998) Functional Food Science and the Cardiovascular System, *British Journal of Nutrition*, vol 80, suppl 1, ppS113–S146
Hosoya, N (1998) Health Claims in Japan, *Japanese Journal of Nutritional Food*, vol 1, no.3/4, pp1–11
Hutt, PB (1986) Government Regulation of Health Claims in Food Labeling and Advertising, *Food, Drug Cosmetic Law Journal*, vol 41, no1, pp3–73
Hutt, PB (1995)A Brief History of FDA Regulation Relating to the Nutrient Content of Food, pp1–27 in Shapiro (1995)
IFIC (International Food Information Council) (1999) see http://ificinfo.health.org
IGD (Institute of Grocery Distributors) (2000) *press release*, 3 August; see www.igd.com
International Association of Consumer Food Organizations (1999) *Functional Foods – Public Health Boon or 21 st Century Quackery?*, Washington, DC, IACFO
Ippolito, P and Mathios, A (1990a) Information, Advertising and Health Choices: A Study of the Cereal Market, *RAND Journal of Economics*, vol 21, no 3, pp459–480
Ippolito, P and Mathios, A (1990b) The Regulation of Science-Based Claims in Advertising, *Journal of Consumer Policy*, vol 13, pp413–45
Ippolito, P (1999) How government policies shape the food and nutrition information environment, *Food Policy*, vol 24, pp295–306
Jackson, R (2000) Guidelines on Preventing Cardiovascular Disease in Clinical Practice, *British Medical Journal*, vol 320, pp659–661
Jacobson, M and Silverglade, B (1999) Functional Foods: Health Boon or Quackery? *British Medical Journal*, vol 319, pp205–206
James, WPT (1988) Dietary Reform: an Individual or National Response, *RSA Journal*, vol CXXXVI, no 5382, pp373–397
James, WPT (1995) A Public Health Approach to the Problem of Obesity, *International Journal of Obesity,* vol 19, suppl 3, ppS37–S45

James, WPT (1997a) *Nutrition in the Future: Thinking the unthinkable*, London, Caroline Walker Trust
James, WPT (1997b) Where Do We Go From Here in Public Health? pp281–292 in Shetty and McPherson (1997)
James, WPT, Nelson, M, Ralph, A and Leather, S (1997a) Socioeconomic Determinants of Health: The Contribution of Nutrition to Inequalities in Health, *British Medical Journal*, vol 314, no 7093, pp1545–1549
James, WPT, Ralph, A, and Bellizzi, M (1997b) Nutrition Policy in Western Europe: National Policies in Belgium, the Netherlands, France, Ireland, and the United Kingdom, *Nutrition Reviews*, vol 55, no 11, ppS4–S20
Jensen, R (1999) *The Dream Society*, New York, McGraw-Hill
JHFNFA (1999) Association documentation and personal visit, May
Kailasapathy, K and Rybka, S (1997) *L. acidophilus* and *Bifodobacterium spp* – Their Therapeutic Potential and Survival in Yogurt, *The Australian Journal of Dairy Technology*, vol 52, pp28–35
Karpf, A (1988) *Doctoring the Media: The Reporting of Health and Medicine*, London, Routledge
Keane, A (1997) Too Hard to Swallow. The Palatability of Healthy Eating Advice, pp172–192 in Caplan (1997)
Keys, A (1953) Atherosclerosis: a Problem in New Public Health, *Journal of Mt Sinai Hospital*, vol 20, pp118–139
Keys, A (1970) *Coronary Heart Disease in Seven Countries*, New York, American Heart Association
Keys, A (1980) *Seven Countries: A Multivariate Analysis of Death and Coronary Heart Disease*, Cambridge, Harvard University Press
Knorr, D (1998) Special Issue: Functional Food Science in Europe, *Trends in Food Science and Technology*, vol 9, pp293–344
Knutson, R, Penn, J, and Boehm, W (1990) *Agricultural and Food Policy*, 2nd edition, New Jersey, Prentice-Hall Inc
Krail, K (2000) *Advertising Age*, February
Kritchevsky, D (1998) 'History of recommendations to the public about dietary fat', paper presented at symposium 'Evolution of ideas about the nutritional value of dietary fat', New Orleans, LA, American Society for Nutritional Sciences, April 9, 1997
Kujala, T (1999) *Rye: Nutrition, Health and Functionality*, Helsinki: Nordic Rye Research Group
Kuulasmaa, K, Tunstall-Pedoe, H, Dobson, A, Fortmann, S, Sans, S, Tolonen, H, Evans, A, Ferrario, M, and Tuomilehto, J (2000) Estimation of Contribution of Changes in Classic Risk Factors to Trends in Coronary-Event Rates Across the WHO MONICA Project Populations, *The Lancet*, vol 335, pp675–687
Lands, W, Hamazaki, T, Yamazaki, K, Okuyama, H, Sakai, K, Goto, Y, and Hubbard, V (1990) Changing dietary patterns, *American Journal of Clinical Nutrition*, vol 51, pp991–993
Lang, T (1997) Going Public: Food Campaigns During the 1980s and Early 1990s, pp238–260 in Smith (1997)

Lang, T and Heasman, M (2001) *Food Wars: The battle for mouths, minds and markets*, London, Earthscan

Lawrence, M and Rayner, M (1998) Functional Foods and Health Claims: A Public Health Policy Perspective, *Public Health Nutrition*, vol 1, no 2, pp75-82

Leather, S (1996) *The Making of Modern Malnutrition*, London, Caroline Walker Lecture 1996

Leather, S (1996) *The Making of Modern Malnutrition. An Overview of Food Poverty in the UK*, Caroline Walker Lecture 1996, London, The Caroline Walker Trust

Lee, Y and Salminen, S (1995) The Coming of Age of Probiotics, *Trends in Food Science and Technology*, vol 6, pp241–245

Levine, M, Rumsey, S, Daruwala, R, Park, J, and Wang, Y (1999) Criteria and Recommendations for Vitamin C Intake, *JAMA*, vol 281, no 15, pp1415–1423

Lilly, D and Stillwell, R (1965) Probiotics: Growth Promoting Factors Produced by Microorganisms, *Science*, vol 147, pp747–748

Lupton, D (1999) *Risk*, London, Routledge

McIntosh, A, Torres, C, Wolinsky, F, Kubena, K, and Landmann, W (1990) Nutritional Risk and Social Relationships, paper presented at the Association For the Study of Food and Society, Fourth Annual Meeting, Philadelphia

McKinlay, J and Marceau, L (2000) Public Health Matters. To Boldly Go..., *American Journal of Public Health*, vol 90, no 1, pp25–33

McColl, D (2000) *Regulatory Aspects of Health-Related Claims*, paper presented at the IFT Short Course on Functional Foods, Orlando, Florida, 29–31 March 2000

McNamara, S (2000) *Functional Foods: Regulatory Issues*, paper given at AACC Conference Functional Foods – Strategies for the Food Industry, Seattle, Washington, 21–22 August 2000

McNutt, K (1980) Dietary advice to the public, *Nutrition Reviews*, vol 38, pp353–360

Maibach, E and Parrott, R (1995) (eds) *Designing Health Messages: Approaches from Communication Theory and Public Health Practice*, London, Sage

Matsumura, M and Ryley, J (1991) Thirty foods a day, *BNF Nutrition Bulletin*, vol 16, pp83–101

Mazza, G (ed) (1998) *Functional Foods. Biochemical and Processing Aspects*, Lancaster, Pennsylvania, ALT Press

Meleady, R and Graham, I (1999) Plasma Homocysteine as a Cardiovascular Risk Factor: Causal, Consequential, or of No Consequence? *Nutrition Reviews*, vol 57, no 10, pp299–303

Miettinen, T, Puska, P, Gylling, H, Vanhanen, H, and Vartiainen, E (1995) Reduction of Serum Cholesterol with Sitostanol-Ester Margarine in a Mildly Hypercholesterolaemic Population, *New England Journal of Medicine*, vol 333, pp1308–1312

Milch Marketing, April 1998

Milio, N (1990) Towards Healthy Public Policy in Food and Nutrition, *Public Health*, vol 104, pp45–54
Milio, N (1991) Health, Nutrition, and Public Policy, *Nursing Outlook*, vol 39, no 1, pp6–9
Milio, N (1993) After the Big Bang: Structure of Food and Nutrition Policy-Making Processes in Europe, in van der Heiji, DG, Lowik, M and Ockhuizen, T (eds) (1993) *Food and Nutrition Policy In Europe*, Wageningen, Pudoc Scientific Publishers
Milio, N and Helsing, E (eds) (1998) *European Food and Nutrition Policies in Action*, WHO Regional Publications European Series, no 73, Copenhagen, WHO Regional Office for Europe
Mills, P, Beeson, W, Phillips, R, and Fraser, G (1989) Cohort Study of Diet, Lifestyle, and Prostate Cancer in Adventist Men, *Cancer*, vol 64, pp598–604
Milner, JA (1999) Functional Foods and Health Promotion, *The Journal of Nutrition*, vol 129, no 7S, suppl, ppS1395–S1397
Mintel (1999) *Functional Foods*, report published April, London, Mintel
Mitchell, V-W (1999) Consumer Perceived Risk: Conceptualisations and Models, *European Journal of Marketing*, vol 33, no 1, pp163–195
NACNE (National Advisory Committee on Nutrition Education) (1983) *A Discussion Paper on Proposals for Nutritional Guidelines for Health Education in Britain*, London, Health Education Council
National Food Authority (1994) *Functional Foods Policy Discussion Paper*, Canberra, National Food Authority
National Public Health Institute (1999) *Nutrition in Finland*, Helsinki: National Public Health Institute
National Food Alliance (1995) Functional Foods: First Look from a Consumer Perspective, London
National Food Alliance (1998) Draft Code of Practice on Health Claims on Foods, London
National Food Authority (1994) Functional Foods: Policy Discussion Paper, Canberra
NCC (1997) *Messages On Food: Consumers Use and Understanding of Health Claims on Food Packs*, London, National Consumer Council
Nestle, M (1993) Dietary advice for the 1990s: the political history of the food guide pyramid, *Caduceus*, vol 9, pp136–153
Nestle, M (1994) Editorial: The Politics of Dietary Guidelines – A New Opportunity, *American Journal of Public Health*, vol 84, no 5, pp713–714
Nestle, M (1999) Commentary, *Food Policy*, vol 24, pp307–310
Norum, K (1997) Some Aspects of Norwegian Nutrition and Food Policy, pp195–206 in Shetty and McPherson (1997)
The Nutraceuticals Institute (1998) *Advancing Nutraceuticals Opportunities: Priorities for Research*, Rutgers, State University of New Jersey
New Nutrition Business (1999 and 2000) the reader is referred to back issues of *New Nutrition Business* (vol 3, October 1998–September 1999 and vol 4, October 1999–March 2000 for market figures referred

to in this book. Current issues update the market stories detailed in this book, see also: www.new-nutrition.com)

Nutrition and Your Health: Dietary Guidelines for Americans (1980) US Department of Agriculture and US Department of Health, Education and Welfare, Home and Garden Bulletin No 232, Washington, DC, US Department of Agriculture

OECD (1981) *Food Policy,* Paris

Omenn G, Goodman, G, Thornquist, M, Balmes, J, Cullen, M, Glass, A, Keogh, J, Meyskens, F, Valanis, B, Williams, J, Barnhardt, S, and Hammar, S (1996) Effects of a combination of beta carotene and vitamin A on lung cancer and cardiovascular disease, *New England Journal of Medicine*, vol 334, pp1150–1155

Ovesen, L (1999) Some pertinent aspects of functional foods, in Council of Europe (1999) *Forum on Functional Food*, Strasbourg, Council of Europe, pp263–288

Oxman, A., Cook, D., and Guyatt, G.(1994) Users' Guide to the Medical Literature, *JAMA*, vol 272, no 17, pp1367–1371

PA Consulting Group (1990) *Functional Foods: A New Global Added Value Market?* London

Passmore, R (1985)Food Propagandists – the New Puritans, *Journal of the Royal College of General Practitioners*, vol 35, pp387–389

Paul, G, Ink, S, and Geiger, C (1999) The Quaker Oats Health Claim: A Case Study, *Journal of Nutraceuticals, Functional Foods & Medical Foods*, vol 1, no 4, pp5–32

Pietinen, P, Rimm, E, Korhonen, P, Hartman, A, Willett, W, Albanes, D, and Virtamo, J (1996) Changes in diet in Finland from 1972 to 1992: Impact on coronary heart risk, *Preventive Medicine*, vol 25, pp243–250

Popcorn, F (1991) *The Popcorn Report*, London, Arrow Books

Porter, M (1990) *The Competitive Advantage of Nations*, New York, Free Press

Porter, M (1998a) Clusters and the new economics of competition, *Harvard Business Review*, November/December, pp77–90

Porter, M (1998b) *On Competition*, Boston, Harvard Business School Press

Poutanen, K (2000) *personal communication*; contact VTT Biotechnology at www.inf.vtt.fi

Powell, D and Leiss, W (1997): *Mad Cows and Mother's Milk: The Perils of Poor Risk Communication*, Montreal, McGill-Queen's University Press

Psaty, B, Weiss, N, Furberg, C, Koepsell, T, Siscovick, D, Rosendaal, F, Smith, N, Heckbert, S, Kaplan, R, Lin, D, Fleming, T, and Wagner, E (1999) Surrogate End Points, Health Outcomes, and the Drug-Approval Process for the Treatment of Risk Factors for Cardiovascular Disease, *JAMA*, vol 282, no 8, pp786-790

Puska, P, Tuomilehto, J, Nissinen, A, and Vartiainen, E (1995) (eds) *The North Karelia Project: 20 Year Results and Experiences*, Finland, Helsinki, National Public Health Institute (KTL)

Raisio Goup plc web-site: www.raisiogroup.com

Rao, A and Agarwal, S (1999) Role of Lycopene as Antioxidant Carotenoid in the Prevention of Chronic Diseases: A Review, *Nutrition Research*, vol 19, no 2, pp305–323

Rayner, M (1998) Systematic Review as a Method for Assessing the Validity of Health Claims, pp174–183 in Sadler and Saltmarsh (1998)

Richardson, DP (1998) Scientific and Regulatory Issues About Food which Claim to Have a Positive Effect on Health, pp196–208 in Sadler and Saltmarsh (1998)

Rifkin, J (2000) *The Age of Access*, New York, Penguin Putnam Inc

Roberfroid, M (1998) Prebiotics and Synbiotics: Concepts and Nutritional Properties, *British Journal of Nutrition*, vol 80, suppl 2, ppS197–S202

Ritzer, G (1992) *McDonaldization of Society*, Thousand Oaks, CA, Sage

Robson, J et al (2000) Estimating Cardiovascular Risk for Primary Prevention: Outstanding Questions for Primary Care, *British Medical Journal*, vol 320, pp702–704

Rychlik, K and Greenwald, G (1997) Market Opportunities: Ingredients, Healthy Foods and Nutraceuticals, pp13–36 in Yalpani (1997)

Sadler, M and Saltmarsh, M (ed) (1998) *Functional Foods. The Consumer, The Products, The Evidence*, London, Royal Society of Chemistry

Salminen, S, Bouley, C, Boutron-Ruault, M, Cummings, J, Franck, A, Gibson, G, Isolauri, E, Moreau, M, Roberfroid, M, and Rowland, I (1998a) Functional Food Science and Gastrointestinal Physiology and Function, *British Journal of Nutrition*, vol 80, suppl 1, ppS147–S171

Salminen, S, Ouwehand, A, and Isolauri, E (1998b) Clinical Applications of Probiotic Bacteria, *International Dairy Journal*, vol 8 pp563–572

Salminen, S and Wright, von A (eds) (1998) *Lactic Acid Bacteria. Microbiology and Functional Aspects*, 2nd edition, New York, Marcel Dekker Inc

Sanders, T (1999) Food Production and Food Safety, *British Medical Journal*, vol 318, pp1689–1693

Scrimshaw, N (1990) Nutrition: Prospects for the 1990s, *Annual Review of Public Health*, vol 11, pp53–68

Shapiro, R (ed) (1995) *Nutrition Labeling Handbook*, New York, Marcel Dekker Inc

Shetty, P and McPherson, K (eds) (1997) *Diet, Nutrition & Chronic Disease. Lessons from Contrasting Worlds*, Chichester, John Wiley & Sons

Silverglade, B (1997) Using Food Labeling to Improve Diet and Health, *European Food Law Review*, vol 8, no 4, pp430–436

Sims, L (1998) *The Politics of Fat. Food and Nutrition Policy in America*, New York, ME Sharpe Inc

Skolbekken, J-A (1995) The Risk Epidemic in Medical Journals, *Social Science and Medicine*, vol 40, pp291–305

Sloan, E (1994) Top ten trends to watch and work on, *Food Technology*, vol 48, pp89–100

Smith, D (ed) (1997) *Nutrition in Britain. Science, Scientists and Politics in the Twentieth Century*, London, Routledge

Stayner, L (1999) Protecting Public Health in the Face of Uncertain Risks: The Example of Diesel Exhaust, *American Journal of Public Health*, vol 89, no 7, pp991–993

Steiner, J (1999) Talking About Treatment: The Language of Populations and the Language of Individuals, *Annals of Internal Medicine*, vol 130, no 7, pp618–622

Stein, AD (1995) Annotation: Cause and Noncause – Nutritional Epidemiology and Public Health Nutrition, *American Journal of Public Health*, vol 85, no 5, pp618–620

Stephen, AM (1998) Regulatory Aspects of Functional Foods, pp403–437 in Mazza (1998)

Swedish Nutrition Foundation (1996) Health Claims in the Labelling and Marketing of Food Products – Revised Self-Regulation Programme, 28 August, Lund, Swedish Nutrition Foundation

Tate & Lyle (1999) *The Sucralose Report*, London: Tate & Lyle

Temple, N (1999) Survey of Nutrition Knowledge of Canadian Physicians, *Journal of the American College of Nutrition*, vol 18, no 1, pp26–29

Temple, R (1999) Are Surrogate Markers Adequate to Assess Cardiovascular Disease Drugs? *JAMA*, vol 282, no 8, pp790–795

Truswell, AS (1987) Evolution of Dietary Recommendations, Goals and Guidelines, *American Journal of Clinical Nutrition*, vol 45, pp1060–1072

Truswell, AS (1998) Practical and Realistic Approaches to Healthier Diet Modifications, *American Journal of Clinical Nutrition*, vol 67, suppl pp583S–590S

UNICEF (1998) *State of the World's Children*, New York, UN Children's Fund/Oxford University Press

Unilever Research Laboratorium (1998) *Diet and Health News*, no 2, The Netherlands, Unilever Nutrition Centre

Valsta, L (1999) Food-based dietary guidelines for Finland – a staged approach, *British Journal of Nutrition*, vol 81, Suppl 2, ppS49–S55

Wearne, SJ and Day, M (1999) Clues for the Development of Food-Based Dietary Guidelines: How are Dietary Targets Being Achieved by UK Consumers?, *British Journal of Nutrition*, vol 81, suppl 2, ppS119–S126

Wheelock, V (1986) *The Food Revolution*, Marlow, Chalcombe Publications

Wheelock, V (1992) Healthy Eating: The Food Issue of the 1990s, *British Food Journal*, vol 94, no 2, pp3–8

White, C (1998) Food is a Public Health Issue, *British Medical Journal*, vol 316, pp416

Willett, WC (1994) Diet and Health: What Should We Eat? *Science*, vol 264, pp 532–537

Willett, W (1997) Dietary fats and non-communicable diseases, in Shetty, P and McPherson, K (eds) *Diet, Nutrition and Chronic Disease:*

Lessons from Contrasting Worlds, Chichester, John Wiley & Sons, pp99–117
Williams, S and Calnan, M (1996) The 'Limits' of Medicalization?: Modern Medicine and the Lay Populace of 'Late' Modernity, *Social Science and Medicine*, vol 42, no 12, pp1609–1620
Winkler, J (1996) Functional Foods: the challenges for consumer policy, *Consumer Policy Review*, vol 6, no 6, pp210–214
Winkler, JT (1998) The Future of Functional Foods, pp184–195 in Sadler and Saltmarsh (1998)
Wiseman, MJ (1990) Government: Where Does Nutrition Policy Come From? *Proceedings of the Nutrition Society*, vol 49, pp397–401
Witwer, R (1998) *Roadmaps to Market, Commercializing Functional Foods and Nutraceuticals*, Waltham, Massachusetts, Decision Resources
Wrick, K (1995) Consumer issues and expectations for functional foods, *Critical Reviews in Food Science and Nutrition*, vol 35, no 1 and 2, pp167–173
Yalpani, M (ed) (1997) *New Technologies for Healthy Foods and Nutraceuticals*, Shrewsbury, MA, ATL Press

The reader is referred to *New Nutrition Business*, Volume 3, October 1998– September 1999 (11 issues) and Volume 4, October 1999–March 2000 (five issues)

Index

Page numbers in *italics* refer to boxes, figures and tables